Common Core Grade 5 Math Workbook

The Most Effective Exercises and Review Common Core Math Questions

By
Elise Baniam & Michael Smith

Common Core Math Workbook

Common Core Grade 5 Math Workbook
Published in the United State of America By
The Math Notion
Email: info@Mathnotion.com
Web: WWW.MathNotion.com

Copyright © 2020 by the Math Notion. All rights reserved. No part of this publication may be reproduced, stored in a retrieval system, or transmitted in any form or by any means, electronic, mechanical, photocopying, recording, scanning, or otherwise, except as permitted under Section 107 or 108 of the 1976 United States Copyright Ac, without permission of the author.
All inquiries should be addressed to the Math Notion.

About the Author

Elise Baniam has been a math instructor for over a decade now. She graduated in Mathematics. Since 2006, Elise has devoted his time to both teaching and developing exceptional math learning materials. As a Math instructor and test prep expert, Elise has worked with thousands of students. She has used the feedback of her students to develop a unique study program that can be used by students to drastically improve their math score fast and effectively.

– **SAT Math Workbook**

– **ACT Math Workbook**

– **PSAT Math Workbooks**

– **ISEE Math Workbooks**

– **SSAT Math Workbooks**

–**many Math Education Workbooks and Practice Books**

– **and some Mathematics books …**

As an experienced Math teacher, Mrs. Baniam employs a variety of formats to help students achieve their goals: she teaches students in large groups, and she provides training materials and textbooks through her website and through Amazon.

You can contact Elise via email at:

Elise@Mathnotion.com

Common Core Grade 5 Math Workbook

This authoritative Common Core Math Workbook makes learning math simple and fun. This updated Common Core Exercises reflects the latest updates to help you achieve the next level of professional achievement.

This prep exercise book and features gives you that edge you need to be successful on Common Core Math Exam. The Common Core Math Workbook covers:

- Number operations/number sense
- Fractions and Mix Numbers
- Algebra, Patterns and Measurements
- Geometry, Symmetry, Data and Graphs

This user-friendly resource includes simple explanations:

- ✓ Review thorough breakdown questions of the Common Core math test
- ✓ **2,000+ Realistic Common Core Math Practice Questions** with answers
- ✓ The Most Feared Subject Made Easier
- ✓ Detailed subjects review, an extensive subject list to help you build your math knowledge
- ✓ **Two Full-length Common Core Practice Tests** with detailed explanations for review and study
- ✓ Help test-taker recognize and pinpoint areas to produce better results in less time
- ✓ Common Core Prep Exams to hone your test-taking techniques

Anyone who wants to realize the major subjects and subtle guidelines of Common Core Math Test, The Common Core Math Workbook offers comprehensive, straightforward instruction.

GET THE ALL-IN-ONE SOLUTION FOR YOUR HIGHEST POSSIBLE 5th Grade Common Core MATH SCORE (Including 2 full-length practice tests for realistic prep, content reviews for math test sections).

WWW.MathNotion.com

WWW.MathNotion.com

… So Much More Online!

✓ FREE Math Lessons

✓ More Math Learning Books!

✓ Mathematics Worksheets

✓ Online Math Tutors

For a PDF Version of This Book

Please Visit WWW.MathNotion.com

Contents

Chapter 1: Place Value and Number Sense ... 11
- Numbers in Standard Form ... 12
- Number in Expand Form ... 13
- Odd or Even ... 14
- Compare Whole Numbers ... 15
- Pattern ... 16
- Round Whole Numbers ... 17
- Answer key Chapter 1 ... 18

Chapter 2: Whole Number Operations ... 21
- Order of Operations ... 22
- Estimate Sums ... 23
- Estimate Differences ... 24
- Subtract from Whole Thousands ... 25
- Multiplication Whole Number ... 26
- Long Division by Two Digit ... 27
- Division with Remainders ... 27
- Dividing Hundreds ... 28
- Answer key Chapter 2 ... 29

Chapter 3: Number Theory ... 31
- Factoring ... 32
- Prime Factorization ... 33
- Divisibility Rule ... 34
- Great Common Factor (GCF) ... 35
- Least Common Multiple (LCM) ... 36
- Answer key Chapter 3 ... 37

Chapter 4: Fractions ... 39
- Adding Fractions – Like Denominator ... 40
- Adding Fractions – Unlike Denominator ... 41
- Subtracting Fractions – Like Denominator ... 42
- Subtracting Fractions – Unlike Denominator ... 43
- Converting Mix Numbers ... 44

Converting improper Fractions 45
Adding Mix Numbers 46
Subtracting Mix Numbers 47
Simplify Fractions.. 48
Multiplying Fractions... 49
Multiplying Mixed Number 50
Dividing Fractions 51
Dividing Mixed Number 52
Comparing Fractions 53
Answer key Chapter 4... 54

Chapter 5: Decimal 59
Graph Decimals... 60
Round Decimals.. 61
Decimals Addition 62
Decimals Subtraction 63
Decimals Multiplication 64
Decimal Division 65
Comparing Decimals 66
Convert Fraction to Decimal 67
Answer key Chapter 5... 68

Chapter 6: Exponent and Radicals.. 71
Positive Exponents 72
Negative Exponents 73
Add and subtract Exponents 74
Exponent multiplication 75
Exponent division 76
Scientific Notation 77
Square Roots 78
Simplify Square Roots 79
Answer key Chapter 6... 80

Chapter 7: Ratio, Proportion and Percent 83
Proportions.. 84
Reduce Ratio 85
Word Problems 86

Percent 87
Convert Fraction to Percent 88
Convert Decimal to Percent 89
Answer key Chapter 7..90

Chapter 8: Measurement 93
Reference Measurement...94
Metric Length Measurement 95
Customary Length Measurement 95
Metric Capacity Measurement 96
Customary Capacity Measurement 96
Metric Weight and Mass Measurement 97
Customary Weight and Mass Measurement...97
Time 98
Answers of Worksheets – Chapter 8...99

Chapter 9: Algebraic Expressions 101
Find a Rule..102
Variables and Expressions..103
Translate Phrases 104
Distributive Property 105
Evaluate One Variable Expressions 106
Answer key Chapter 9..107

Chapter 10: Symmetry and Transformations 109
Line Segments 110
Parallel, Perpendicular and Intersecting Lines 111
Identify Lines of Symmetry 112
Lines of Symmetry 113
Identify Three–Dimensional Figures...114
Vertices, Edges, and Faces 115
Identify Faces of Three–Dimensional Figures 116
Answers of Worksheets – Chapter 10..117

Chapter 11: Geometry 121
Area and Perimeter of Square 122
Area and Perimeter of Rectangle 123
Area and Perimeter of Triangle..124

Area and Perimeter of Trapezoid 125
Area and Perimeter of Parallelogram 126
Circumference and Area of Circle 127
Perimeter of Polygon 128
Volume of Cubes 129
Volume of Rectangle Prism 130
Answer key Chapter 11 131

Chapter 12: Data and Graphs 133
Mean and Median 134
Mode and Range 135
Stem–And–Leaf Plot 136
Dot plots 137
Bar Graph 138
Probability 139
Answer key Chapter 12 140

Common Core Test Review 143
Common Core Practice Test 1 147
Common Core Practice Test 2 161

Answers and Explanations 175
Answer Key 177
Common Core Practice Test 1 179
Common Core Practice Test 2 185

Chapter 1: Place Value and Number Sense

Numbers in Standard Form

Write the number in standard form.

1) 14 million 154 thousand 8

2) 89 million 15 thousand 798

3) 97 million 5 thousand 8

4) 124 million 2 thousand 2

5) 50 billion 3 million 5 thousand 4

6) 34 billion 45 million 578

7) 94 billion 21 million 51 thousand

8) 58 billion 708 thousand 120

9) 59 billion 54 thousand 86

10) 74 billion 354 thousand 158

11) 7 billion 13 million 12 thousand 7

12) 72 billion 450 million 658

13) 398 million 67 thousand 128

14) 24 billion 54 million 9 thousand 32

15) 795 million 458

16) 38 billion 2 million 54 thousand 9

Number in Expand Form

Write the number in expand form.

1) 788: _____.

2) 1,200: _____.

3) 77,153: _____.

4) 80,050: _____.

5) 302,186: _____.

6) 58,604: _____.

7) 916,236: _____.

8) 3,625: _____.

9) 45,029: _____.

10) 66,312: _____.

11) 7,293: _____.

12) 500,538: _____.

13) 900,900: _____.

14) 1,040,000: _____.

Odd or Even

Write odd or even.

1) 17 _____

2) 65 _____

3) 584 _____

4) 468 _____

5) 769 _____

6) 123 _____

7) 701 _____

8) 140 _____

9) 999 _____

10) 816 _____

11) 21 _____

12) 5,236 _____

13) 12,457 _____

14) 6,589 _____

15) 1,000 _____

16) 18 _____

17) 542 _____

18) 450 _____

19) 56,189 _____

20) 18,200 _____

21) 168 _____

22) 65,231 _____

23) 56 _____

24) 6,900 _____

25) 1,659 _____

26) 867 _____

Compare Whole Numbers

Compare, writing <, >, or = between the numbers.

1) 50,850 ☐ 57,580

2) 38,380 ☐ 38,830

3) 79,970 ☐ 97,790

4) 55,750 ☐ 55,570

5) 68,324 ☐ 67,820

6) 75,862 ☐ 75,860

7) 98,687 ☐ 98,876

8) 82,857 ☐ 82,758

9) 63,454 ☐ 63,545

10) 55,636 ☐ 56,636

11) 24,880 ☐ 42,880

12) 98,898 ☐ 89,899

13) 59,350 ☐ 85,220

14) 71,970 ☐ 71,870

15) 38,680 ☐ 38,860

16) 64,550 ☐ 64,550

17) 121,980 ☐ 121,890

18) 109,807 ☐ 109,807

19) 154,150 ☐ 154,510

20) 245,250 ☐ 254,250

21) 123,459 ☐ 132,945

22) 145,753 ☐ 145,375

23) 265,789 ☐ 246,987

24) 186,896 ☐ 186,689

25) 149,368 ☐ 151,120

26) 201,578 ☐ 199,990

27) 215,887 ☐ 251,887

28) 301,980 ☐ 301,890

Common Core Math Workbook

Pattern

Continue this pattern for four more numbers:

1) 1,500; 1,350; 1,200; 1,050; _____

2) 2,800; 2,600; 2,400; 2,200; _____

3) 3,500; 3,150; 2,800; 2,450; _____

4) 1,900; 1,780; 1,660; 1,540; _____

5) 3,200; 2,950; 2,700; 2,450; _____

6) 4,100; 3,800; 3,500; 3,200; _____

7) 5,400; 4,950; 4,500; 4,050; _____

8) 2,900; 2,725; 2,550; 2,375; _____

9) 1,950; 1,700; 1,450; 1,200; _____

10) 5,500; 4,900; 4,300; 3,700; _____

11) Write a list of five numbers that follows this pattern: Start at 100 and add 400 each time.

_____

WWW.MathNotion.com

Round Whole Numbers

Round to the place of the underlined digit.

1) 8,359,536 ≈ _____

2) 978,245 ≈ _____

3) 5,324,305 ≈ _____

4) 11,246,785 ≈ _____

5) 7,366,552 ≈ _____

6) 4,467,859 ≈ _____

7) 3,458,724 ≈ _____

8) 10,320,668 ≈ _____

9) 12,120,809 ≈ _____

10) 9,935,890 ≈ _____

11) 10,327,758 ≈ _____

12) 3,147,904 ≈ _____

13) 7,394,885 ≈ _____

14) 5,358,568 ≈ _____

15) 4,386,709 ≈ _____

16) 9,555,665 ≈ _____

17) 2,324,012 ≈ _____

18) 7,425,594 ≈ _____

19) 11,167,780 ≈ _____

20) 9,338,421 ≈ _____

21) 1,846,102 ≈ _____

22) 9,324,489 ≈ _____

Answer key Chapter 1

Numbers in Standard Form

1) 14,154,008
2) 89,015,798
3) 97,005,008
4) 124,002,002
9) 59,000,054,086
10) 74,000,354,158
11) 7,013,012,007
12) 72,450,000,658
5) 50,003,005,004
6) 34,054,000,578
7) 94,021,051,000
8) 58,000,708,120
13) 398,067,128
14) 24,054,009,032
15) 795,000,458
16) 38,002,054,009

Numbers in Expand Form

1) $(7 \times 100) + (8 \times 10) + 8$
2) $(1 \times 1,000) + (2 \times 100)$
3) $(7 \times 10,000) + (7 \times 1,000) + (1 \times 100) + (5 \times 10) + 3$
4) $(8 \times 10,000) + (5 \times 10) + 0$
5) $(3 \times 100,000) + (2 \times 1,000) + (1 \times 100) + (8 \times 10) + 6$
6) $(5 \times 10,000) + (8 \times 1,000) + (6 \times 100) + 4$
7) $(9 \times 100,000) + (1 \times 10,000) + (6 \times 1,000) + (2 \times 100) + (3 \times 10) + 6$
8) $(3 \times 1,000) + (6 \times 100) + (2 \times 10) + 5$
9) $(4 \times 10,000) + (5 \times 1,000) + (2 \times 10) + 9$
10) $(6 \times 10,000) + (6 \times 1,000) + (3 \times 100) + (1 \times 10) + 2$
11) $(7 \times 1,000) + (2 \times 100) + (9 \times 10) + 3$
12) $(5 \times 100,000) + (5 \times 100) + (3 \times 10) + 8$
13) $(9 \times 100,000) + (9 \times 100)$
14) $(1 \times 1,000,000) + (4 \times 10,000)$

Odd or Even

1) Odd
2) Odd
3) Even
4) Even
5) Odd
6) Odd
7) Odd
8) Even
9) Odd
10) Even
11) Odd
12) Even

13) Odd
14) Odd
15) Even
16) Even
17) Odd
18) Even
19) Odd
20) Even
21) Even
22) odd
23) even
24) Even
25) odd
26) odd

Compare Whole Numbers

1) <
2) <
3) <
4) >
5) >
6) >
7) <
8) >
9) <
10) <
11) <
12) >
13) <
14) >
15) <
16) =
17) >
18) =
19) <
20) <
21) <
22) >
23) >
24) >
25) <
26) >
27) <
28) >

Pattern

1) 900; 750; 600; 450
2) 2,000; 1,800; 1,600; 1,400
3) 2,100; 1,750; 1,400; 1,050
4) 1,420; 1,300; 1,180; 1,060
5) 2,200; 1,950; 1,700; 1,450
6) 2,900; 2,600; 2,300; 2,000
7) 3,600; 3,150; 2,700; 2,250
8) 2,200; 2,025; 1,850; 1,675
9) 950; 700; 450; 200
10) 3,100; 2,500; 1,900; 1,300
11) 100; 500; 900; 1,300; 1,700

Round whole number

1) 8,360,000
2) 978,000
3) 5,324,000
4) 11,247,000
5) 7,367,000
6) 4,468,000
7) 3,458,700
8) 10,320,700
9) 12,120,800
10) 9,900,000
11) 10,328,000
12) 3,148,000
13) 7,390,000
14) 5,359,000
15) 4,386,700
16) 9,556,000
17) 2,324,000
18) 7,425,590
19) 1,168,000
20) 9,340,000
21) 1,850,000
22) 9,324,000

Chapter 2: Whole Number Operations

Order of Operations

Evaluate each expression.

1) $6 + (2 \times 5) =$

2) $15 - (4 \times 2) =$

3) $(14 \times 5) + 15 =$

4) $(18 - 3) - (4 \times 5) =$

5) $32 + (18 \div 3) =$

6) $(18 \times 4) \div 6 =$

7) $(63 \div 7) \times (-3) =$

8) $(7 \times 8) + (34 - 18) =$

9) $80 + (3 \times 3) + 5 =$

10) $(20 \times 8) \div (4 + 4) =$

11) $(-7) + (12 \times 5) + 11 =$

12) $(5 \times 9) - (45 \div 5) =$

13) $(7 \times 6 \div 2) - (17 + 13) =$

14) $(14 + 6 - 16) \times 8 - 12 =$

15) $(36 - 18 + 30) \times (96 \div 8) =$

16) $24 + (14 - (36 \div 6)) =$

17) $(7 + 10 - 4 - 9) + (24 \div 3) =$

18) $(90 - 15) + (16 - 18 + 8) =$

19) $(20 \times 3) + (16 \times 4) - 10 =$

20) $15 + 12 - (26 \times 5) + 15 =$

Estimate Sums

Estimate the sum by rounding each added to the nearest ten.

1) $36 + 9 =$

2) $29 + 46 =$

3) $36 + 12 =$

4) $37 + 38 =$

5) $12 + 35 =$

6) $38 + 13 =$

7) $48 + 25 =$

8) $36 + 77 =$

9) $45 + 86 =$

10) $62 + 58 =$

11) $45 + 36 =$

12) $52 + 18 =$

13) $35 + 59 =$

14) $38 + 65 =$

15) $87 + 82 =$

16) $18 + 69 =$

17) $65 + 64 =$

18) $33 + 26 =$

19) $73 + 48 =$

20) $35 + 64 =$

21) $13 + 93 =$

22) $63 + 52 =$

23) $164 + 142 =$

24) $54 + 77 =$

Estimate Differences

Estimate the difference by rounding each number to the nearest ten.

1) $58 - 23 =$

2) $34 - 24 =$

3) $75 - 48 =$

4) $43 - 24 =$

5) $69 - 46 =$

6) $42 - 23 =$

7) $77 - 47 =$

8) $49 - 28 =$

9) $94 - 48 =$

10) $79 - 59 =$

11) $68 - 26 =$

12) $83 - 37 =$

13) $73 - 43 =$

14) $58 - 42 =$

15) $82 - 52 =$

16) $65 - 43 =$

17) $99 - 81 =$

18) $42 - 24 =$

19) $58 - 47 =$

20) $89 - 28 =$

21) $81 - 65 =$

22) $68 - 14 =$

23) $76 - 6 =$

24) $78 - 31 =$

Subtract from Whole Thousands.

Find the sum or difference.

1) 3,000 − 10 = ___

2) 4,000 − 5 = ___

3) 2,000 − 8 = ___

4) 5,000 − 30 = ___

5) 7,000 − 7 = ___

6) 6,000 − 15 = ___

7) 8,000 − 40 = ___

8) 9,000 − 5 = ___

9) 2,000 − 8 = ___

10) 5,000 − 30 = ___

11) 7,000 − 200 = ___

12) 6,000 − 2 = ___

13) 4,000 − 20 = ___

14) 8,000 − 200 = ___

15) 5,000 − 100 = ___

16) 6,000 − 80 = ___

17) 5,000 − 70 = ___

18) 7,000 − 200 = ___

19) 9,000 − 300 = ___

20) 2,000 − 8 = ___

21) 4,000 − 10 = ___

22) 8,000 − 50 = ___

23) 3,000 − 90 = ___

24) 1,000 − 6 = ___

25) 5,000 − 5 = ___

26) 8,000 − 90 = ___

27) 9,000 − 30 = ___

28) 2,000 − 60 = ___

Multiplication Whole Number

Calculate.

1) $210 \times 9 =$

2) $160 \times 40 =$

3) $(-6) \times 8 \times (-5) =$

4) $-5 \times (-7) \times (-7) =$

5) $13 \times (-13) =$

6) $40 \times (-8) =$

7) $8 \times (-2) \times 6 =$

8) $(-400) \times (-30) =$

9) $(-20) \times (-20) \times 3 =$

10) $125 \times 6 =$

11) $142 \times 50 =$

12) $364 \div 14 =$

13) $(-4{,}125) \div 5 =$

14) $(-28) \div (-7) =$

15) $288 \div (-18) =$

16) $3{,}500 \div 28 =$

17) $(-126) \div 3 =$

18) $4{,}128 \div 4 =$

19) $1{,}260 \div (-35) =$

20) $3{,}360 \div 4 =$

21) $(-54) \div 2 =$

22) $(-2{,}000) \div (-20) =$

23) $0 \div 870 =$

24) $(-1{,}020) \div 6 =$

25) $5{,}868 \div 652 =$

26) $(-2{,}520) \div 4 =$

27) $10{,}902 \div 3 =$

28) $(-60) \div (-5) =$

Common Core Math Workbook

Long Division by Two Digit

Find the quotient.

1) 16)512

2) 12)816

3) 24)672

4) 28)364

5) 34)578

6) 36)324

7) 21)651

8) 42)2,142

9) 65)1,300

10) 45)1,620

11) 63)2,961

12) 50)2,400

13) 27)2,457

14) 67)7,303

15) 93)4,092

16) 76)6,156

17) 70)12,880

18) 18)11,088

Division with Remainders

Find the quotient with remainder.

1) 12)613

2) 15)2,579

3) 23)3,923

4) 81)3,566

5) 38)6,996

6) 75)8,009

7) 59)7,512

8) 85)11,264

9) 45)7,335

10) 88)12,589

11) 36)9,564

12) 60)36,947

13) 78)6,298

14) 95)37,456

Dividing Hundreds

Find answers.

1) $2000 \div 200$

2) $1600 \div 20$

3) $900 \div 100$

4) $3,200 \div 800$

5) $4,800 \div 800$

6) $900 \div 300$

7) $2,400 \div 800$

8) $4,500 \div 900$

9) $6,800 \div 200$

10) $10,000 \div 200$

11) $8,100 \div 300$

12) $8,000 \div 500$

13) $1,200 \div 200$

14) $6,600 \div 600$

15) $7,200 \div 600$

16) $1,800 \div 200$

17) $27,000 \div 900$

18) $9,900 \div 300$

19) $7,200 \div 100$

20) $9,000 \div 120$

21) $9,000 \div 3,000$

22) $16,000 \div 40$

23) $210 \div 30$

24) $560 \div 70$

Answer key Chapter 2

Order of Operations

1) 16
2) 7
3) 85
4) −5
5) 38
6) 12
7) −27
8) 72
9) 94
10) 20
11) 64
12) 36
13) −9
14) 20
15) 576
16) 32
17) 12
18) 81
19) 114
20) −88

Estimate sums

1) 50
2) 80
3) 50
4) 80
5) 50
6) 50
7) 80
8) 120
9) 140
10) 120
11) 90
12) 70
13) 100
14) 110
15) 170
16) 90
17) 130
18) 60
19) 120
20) 100
21) 100
22) 110
23) 300
24) 130

Estimate differences

1) 40
2) 10
3) 30
4) 20
5) 20
6) 20
7) 30
8) 20
9) 40
10) 20
11) 40
12) 40
13) 30
14) 20
15) 30
16) 30
17) 20
18) 20
19) 10
20) 60
21) 10
22) 60
23) 70
24) 50

Subtract from Whole Thousands

1) 2,990
2) 3,995
3) 1,992
4) 4,970
5) 6,993
6) 5,985
7) 7,960
8) 8,995
9) 1,992
10) 4,970
11) 6,800
12) 5,998
13) 3,980
14) 7,800
15) 4,900
16) 5,920
17) 5,930
18) 6,800
19) 8,700
20) 1,992
21) 3,990
22) 7,950
23) 2,910
24) 994
25) 4,995
26) 7,910
27) 8,970

28) 1,940

Multiplication Whole Number

1) 1,890
2) 6,400
3) 240
4) −245
5) −169
6) −320
7) −96
8) 12,000
9) 1,200
10) 750
11) 7,100
12) 26
13) −825
14) 4
15) −16
16) 125
17) −42
18) 1,032
19) −36
20) 840
21) −27
22) 100
23) 0
24) −170
25) 9
26) −630
27) 3,634
28) 12

Long Division by Two Digit

1) 32
2) 68
3) 28
4) 13
5) 17
6) 9
7) 31
8) 51
9) 20
10) 36
11) 47
12) 48
13) 91
14) 109
15) 44
16) 81
17) 184
18) 616

Division with Remainders

1) 51 R1
2) 171 R14
3) 170 R13
4) 44 R2
5) 184 R4
6) 106 R59
7) 127 R19
8) 132 R44
9) 163 R0
10) 143 R5
11) 265 R24
12) 615 R47
13) 80 R58
14) 394 R26

Dividing Hundreds

1) 10
2) 80
3) 9
4) 4
5) 6
6) 3
7) 3
8) 5
9) 34
10) 50
11) 27
12) 16
13) 6
14) 11
15) 12
16) 9
17) 30
18) 33
19) 75
20) 75
21) 3
22) 400
23) 7
24) 8

Chapter 3:
Number Theory

Factoring

Factor, write prime if prime.

1) 16

2) 82

3) 28

4) 46

5) 62

6) 65

7) 38

8) 10

9) 54

10) 85

11) 45

12) 90

13) 81

14) 55

15) 92

16) 114

17) 86

18) 70

19) 105

20) 80

21) 95

22) 36

23) 84

24) 110

25) 34

26) 98

27) 63

28) 106

Prime Factorization

Factor the following numbers to their prime factors.

1.
   20
  / \
  _____

2.
   33
  / \
  _____

3.
   62
  / \
  _____

4.
   30
  / \
  _____

5.
   14
  / \
  _____

6.
   29
  / \
  _____

7.
   36
  / \
  _____

8.
   48
  / \
  _____

9.
   76
  / \
  _____

10.
   92
  / \
  _____

11.
   89
  / \
  _____

12.
   68
  / \
  _____

Divisibility Rule

Apply the divisibility rules to find the factors of each number.

1) 15   2, 3, 4, 5, 6, 9, 10
2) 124   2, 3, 4, 5, 6, 9, 10
3) 352   2, 3, 4, 5, 6, 9, 10
4) 94   2, 3, 4, 5, 6, 9, 10
5) 241   2, 3, 4, 5, 6, 9, 10
6) 455   2, 3, 4, 5, 6, 9, 10
7) 65   2, 3, 4, 5, 6, 9, 10
8) 320   2, 3, 4, 5, 6, 9, 10
9) 1,134 2, 3, 4, 5, 6, 9, 10
10) 68   2, 3, 4, 5, 6, 9, 10
11) 754 2, 3, 4, 5, 6, 9, 10
12) 148 2, 3, 4, 5, 6, 9, 10

13) 34   2, 3, 4, 5, 6, 9, 10
14) 385   2, 3, 4, 5, 6, 9, 10
15) 915   2, 3, 4, 5, 6, 9, 10
16) 157   2, 3, 4, 5, 6, 9, 10
17) 540   2, 3, 4, 5, 6, 9, 10
18) 340   2, 3, 4, 5, 6, 9, 10
19) 480   2, 3, 4, 5, 6, 9, 10
20) 3,750 2, 3, 4, 5, 6, 9, 10
21) 660   2, 3, 4, 5, 6, 9, 10
22) 286   2, 3, 4, 5, 6, 9, 10
23) 158   2, 3, 4, 5, 6, 9, 10
24) 456   2, 3, 4, 5, 6, 9, 10

Great Common Factor (GCF)

Find the GCF of the numbers.

1) 6, 15

2) 46, 27

3) 48, 58

4) 20, 25

5) 16, 36

6) 32, 42

7) 60, 25

8) 90, 35

9) 72, 9

10) 45, 54

11) 88, 54

12) 35, 70

13) 70, 20

14) 32, 82

15) 48, 96

16) 30, 85

17) 16, 24

18) 80, 100, 40

19) 81, 112

20) 56, 88

21) 10, 5, 25

22) 8, 18, 24

23) 15, 45, 60

24) 51, 33

Least Common Multiple (LCM)

Find the LCM of each.

1) 8, 10

2) 32, 16

3) 5, 10, 15

4) 14, 21

5) 20, 4, 30

6) 25, 5

7) 12, 60, 24

8) 5, 6

9) 13, 26, 54

10) 28, 35

11) 27, 54

12) 110, 22

13) 30, 15, 60

14) 18, 63

15) 40, 8, 5

16) 81, 18

17) 38, 19

18) 22, 44

19) 25, 60

20) 16, 48

21) 17, 10

22) 8, 28

23) 35, 70

24) 21, 6

Answer key Chapter 3

Factoring

1) 1, 2, 4, 8, 16
2) 1, 2, 41, 82
3) 1, 2, 4, 7, 14, 28
4) 1, 2, 23, 46
5) 1, 2, 31, 62
6) 1, 5, 13, 65
7) 1, 2, 19, 38
8) 1, 2, 5, 10
9) 1, 2, 3, 6, 9, 18, 27, 54
10) 1, 5, 17, 85
11) 1, 3, 5, 9, 15, 45
12) 1, 2, 3, 5, 6, 9, 10, 15, 18, 30, 45, 90
13) 1, 3, 9, 27, 81
14) 1, 5, 11, 55
15) 1, 2, 4, 23, 46, 92
16) 1, 2, 3, 6, 19, 38, 57, 114
17) 1, 2, 43, 86
18) 1, 2, 5, 7, 10, 14, 35, 70
19) 1, 3, 5, 7, 15, 21, 35, 105
20) 1, 2, 4, 5, 8, 10, 16, 20, 40, 80
21) 1, 5, 19, 95
22) 1, 2, 3, 4, 6, 9, 12, 18, 36
23) 1, 2, 3, 4, 6, 7, 12, 14, 21, 28, 42, 84
24) 1, 2, 5, 10, 11, 22, 55, 110
25) 1, 2, 17, 34
26) 1, 2, 7, 14, 49, 98
27) 1, 3, 7, 9, 21, 63
28) 1, 2, 53, 106

Prime Factorization

1) $2 \times 2 \times 5$
2) 3×11
3) 2×31
4) $2 \times 3 \times 5$
5) 2×7
6) 29 is a prime number
7) $2 \times 2 \times 3 \times 3$
8) $2 \times 2 \times 2 \times 2 \times 3$
9) $2 \times 2 \times 19$
10) $2 \times 2 \times 23$
11) 89 is a prime number
12) $2 \times 2 \times 17$

Divisibility Rule

1) 15 — 2, <u>3</u>, 4, <u>5</u>, 6, 9, 10
2) 124 — <u>2</u>, 3, <u>4</u>, 5, 6, 9, 10
3) 352 — <u>2</u>, 3, <u>4</u>, 5, 6, 9, 10
4) 94 — <u>2</u>, 3, 4, 5, 6, 9, 10
5) 241 — 2, 3, 4, 5, 6, 9, 10
6) 455 — 2, 3, 4, <u>5</u>, 6, 9, 10
7) 65 — 2, 3, 4, <u>5</u>, 6, 9, 10
8) 320 — <u>2</u>, 3, <u>4</u>, <u>5</u>, 6, 9, 10
9) 1,134 — <u>2</u>, <u>3</u>, 4, 5, <u>6</u>, <u>9</u>, 10
10) 68 — <u>2</u>, 3, <u>4</u>, 5, 6, 9, 10
11) 754 — <u>2</u>, 3, 4, 5, 6, 9, 10
12) 148 — <u>2</u>, 3, <u>4</u>, 5, 6, 9, 10

13) 34	2, 3, 4, 5, 6, 9, 10	19) 480	2, 3, 4, 5, 6, 9, 10
14) 385	2, 3, 4, 5, 6, 9, 10	20) 3,750	2, 3, 4, 5, 6, 9, 10
15) 915	2, 3, 4, 5, 6, 9, 10	21) 660	2, 3, 4, 5, 6, 9, 10
16) 157	2, 3, 4, 5, 6, 9, 10	22) 286	2, 3, 4, 5, 6, 9, 10
17) 540	2, 3, 4, 5, 6, 9, 10	23) 158	2, 3, 4, 5, 6, 9, 10
18) 340	2, 3, 4, 5, 6, 9, 10	24) 456	2, 3, 4, 5, 6, 9, 10

Great Common Factor (GCF)

1) 3
2) 1
3) 2
4) 5
5) 4
6) 2
7) 5
8) 5
9) 9
10) 9
11) 2
12) 35
13) 10
14) 2
15) 48
16) 5
17) 8
18) 20
19) 1
20) 8
21) 5
22) 2
23) 15
24)

Least Common Multiple (LCM)

1) 40
2) 32
3) 30
4) 42
5) 60
6) 5
7) 120
8) 30
9) 702
10) 140
11) 54
12) 110
13) 60
14) 126
15) 40
16) 162
17) 38
18) 44
19) 300
20) 48
21) 170
22) 56
23) 70
24) 42

Chapter 4:
Fractions

Adding Fractions – Like Denominator

Find each sum.

1) $\dfrac{1}{4} + \dfrac{2}{4} =$

2) $\dfrac{2}{5} + \dfrac{1}{5} =$

3) $\dfrac{1}{8} + \dfrac{2}{8} =$

4) $\dfrac{4}{11} + \dfrac{1}{11} =$

5) $\dfrac{4}{21} + \dfrac{1}{21} =$

6) $\dfrac{5}{49} + \dfrac{6}{49} =$

7) $\dfrac{2}{7} + \dfrac{11}{7} =$

8) $\dfrac{1}{15} + \dfrac{3}{15} =$

9) $\dfrac{3}{19} + \dfrac{6}{19} =$

10) $\dfrac{1}{13} + \dfrac{1}{13} =$

11) $\dfrac{1}{5} + \dfrac{1}{5} =$

12) $\dfrac{4}{17} + \dfrac{6}{17} =$

13) $\dfrac{2}{20} + \dfrac{17}{20} =$

14) $\dfrac{4}{25} + \dfrac{7}{25} =$

15) $\dfrac{6}{14} + \dfrac{3}{14} =$

16) $\dfrac{12}{30} + \dfrac{5}{30} =$

17) $\dfrac{1}{9} + \dfrac{1}{9} =$

18) $\dfrac{29}{5} + \dfrac{3}{5} =$

19) $\dfrac{18}{6} + \dfrac{5}{6} =$

20) $\dfrac{25}{37} + \dfrac{11}{37} =$

Adding Fractions – Unlike Denominator

Add the fractions and simplify the answers.

1) $\frac{1}{4}+\frac{2}{3}=$

2) $\frac{4}{5}+\frac{1}{2}=$

3) $\frac{1}{4}+\frac{5}{7}=$

4) $\frac{8}{11}+\frac{1}{2}=$

5) $\frac{7}{18}+\frac{1}{3}=$

6) $\frac{13}{54}+\frac{5}{18}=$

7) $\frac{5}{8}+\frac{1}{6}=$

8) $\frac{3}{10}+\frac{1}{4}=$

9) $\frac{5}{11}+\frac{2}{4}=$

10) $\frac{1}{9}+\frac{4}{7}=$

11) $\frac{5}{18}+\frac{3}{8}=$

12) $\frac{7}{32}+\frac{3}{4}=$

13) $\frac{9}{65}+\frac{2}{5}=$

14) $\frac{8}{63}+\frac{3}{7}=$

15) $\frac{11}{64}+\frac{1}{4}=$

16) $\frac{4}{15}+\frac{2}{5}=$

17) $\frac{4}{7}+\frac{3}{6}=$

18) $\frac{5}{72}+\frac{2}{9}=$

19) $\frac{2}{15}+\frac{1}{25}=$

20) $\frac{5}{12}+\frac{3}{8}=$

21) $\frac{7}{88}+\frac{1}{8}=$

22) $\frac{7}{12}+\frac{2}{5}=$

23) $\frac{3}{72}+\frac{1}{4}=$

24) $\frac{2}{27}+\frac{1}{18}=$

Subtracting Fractions – Like Denominator

Find the difference.

1) $\frac{5}{3} - \frac{2}{3} =$

2) $\frac{5}{8} - \frac{3}{8} =$

3) $\frac{11}{14} - \frac{8}{14} =$

4) $\frac{13}{3} - \frac{7}{3} =$

5) $\frac{15}{17} - \frac{13}{17} =$

6) $\frac{18}{33} - \frac{10}{33} =$

7) $\frac{8}{25} - \frac{2}{25} =$

8) $\frac{17}{27} - \frac{2}{27} =$

9) $\frac{7}{10} - \frac{3}{10} =$

10) $\frac{24}{35} - \frac{4}{35} =$

11) $\frac{11}{5} - \frac{3}{5} =$

12) $\frac{28}{38} - \frac{18}{38} =$

13) $\frac{5}{6} - \frac{1}{6} =$

14) $\frac{22}{43} - \frac{11}{43} =$

15) $\frac{4}{7} - \frac{3}{7} =$

16) $\frac{18}{29} - \frac{15}{29} =$

17) $\frac{4}{5} - \frac{3}{5} =$

18) $\frac{42}{53} - \frac{38}{53} =$

19) $\frac{8}{31} - \frac{3}{31} =$

20) $\frac{32}{39} - \frac{30}{39} =$

21) $\frac{9}{26} - \frac{5}{26} =$

22) $\frac{31}{46} - \frac{27}{46} =$

23) $\frac{25}{48} - \frac{19}{48} =$

24) $\frac{39}{65} - \frac{27}{65} =$

Subtracting Fractions – Unlike Denominator

Solve each problem.

1) $\dfrac{3}{4} - \dfrac{1}{5} =$

2) $\dfrac{2}{3} - \dfrac{1}{4} =$

3) $\dfrac{5}{6} - \dfrac{3}{7} =$

4) $\dfrac{5}{6} - \dfrac{7}{12} =$

5) $\dfrac{6}{7} - \dfrac{3}{14} =$

6) $\dfrac{7}{12} - \dfrac{7}{18} =$

7) $\dfrac{17}{20} - \dfrac{2}{5} =$

8) $\dfrac{2}{3} - \dfrac{1}{16} =$

9) $\dfrac{6}{7} - \dfrac{4}{9} =$

10) $\dfrac{3}{8} - \dfrac{5}{32} =$

11) $\dfrac{5}{7} - \dfrac{4}{35} =$

12) $\dfrac{5}{6} - \dfrac{7}{30} =$

13) $\dfrac{6}{7} - \dfrac{4}{21} =$

14) $\dfrac{5}{3} - \dfrac{8}{15} =$

15) $\dfrac{2}{11} - \dfrac{3}{22} =$

16) $\dfrac{5}{6} - \dfrac{4}{54} =$

17) $\dfrac{7}{24} - \dfrac{7}{32} =$

18) $\dfrac{3}{4} - \dfrac{3}{5} =$

19) $\dfrac{1}{2} - \dfrac{2}{9} =$

20) $\dfrac{2}{3} - \dfrac{6}{11} =$

Converting Mix Numbers

Convert the following mixed numbers into improper fractions.

1) $3\frac{5}{6}=$

2) $5\frac{11}{15}=$

3) $4\frac{1}{3}=$

4) $2\frac{4}{7}=$

5) $7\frac{1}{4}=$

6) $3\frac{19}{21}=$

7) $5\frac{9}{10}=$

8) $4\frac{7}{12}=$

9) $3\frac{10}{11}=$

10) $6\frac{2}{5}=$

11) $8\frac{2}{3}=$

12) $2\frac{11}{12}=$

13) $3\frac{5}{6}=$

14) $4\frac{8}{11}=$

15) $7\frac{1}{4}=$

16) $5\frac{6}{11}=$

17) $8\frac{1}{5}=$

18) $3\frac{7}{12}=$

19) $6\frac{1}{22}=$

20) $3\frac{2}{3}=$

21) $7\frac{4}{5}=$

22) $4\frac{7}{8}=$

23) $6\frac{5}{6}=$

24) $12\frac{9}{10}=$

Common Core Math Workbook

Converting improper Fractions

Convert the following improper fractions into mixed numbers

1) $\frac{62}{14} =$

2) $\frac{98}{37} =$

3) $\frac{49}{17} =$

4) $\frac{57}{23} =$

5) $\frac{71}{16} =$

6) $\frac{137}{42} =$

7) $\frac{120}{33} =$

8) $\frac{26}{5} =$

9) $\frac{33}{19} =$

10) $\frac{13}{2} =$

11) $\frac{39}{4} =$

12) $\frac{210}{65} =$

13) $\frac{76}{64} =$

14) $\frac{18}{7} =$

15) $\frac{110}{13} =$

16) $\frac{49}{4} =$

17) $\frac{122}{9} =$

18) $\frac{61}{12} =$

19) $\frac{37}{6} =$

20) $\frac{28}{9} =$

21) $\frac{5}{4} =$

22) $\frac{79}{13} =$

23) $\frac{41}{8} =$

24) $\frac{64}{7} =$

WWW.MathNotion.com

Adding Mix Numbers

Add the following fractions.

1) $1\frac{1}{5} + 4\frac{2}{5} =$

2) $5\frac{3}{7} + 3\frac{4}{7} =$

3) $2\frac{2}{8} + 3\frac{1}{8} =$

4) $5\frac{5}{8} + 3\frac{1}{2} =$

5) $2\frac{9}{14} + 3\frac{3}{12} =$

6) $6\frac{2}{5} + 3\frac{1}{2} =$

7) $2\frac{8}{27} + 2\frac{2}{18} =$

8) $2\frac{3}{4} + 3\frac{1}{3} =$

9) $4\frac{5}{6} + 1\frac{1}{6} =$

10) $3\frac{5}{7} + 1\frac{3}{7} =$

11) $4\frac{1}{2} + 2\frac{2}{5} =$

12) $5\frac{1}{4} + 2\frac{5}{6} =$

13) $5\frac{1}{3} + 2\frac{2}{3} =$

14) $3\frac{5}{6} + 3\frac{2}{12} =$

15) $4\frac{3}{5} + 4\frac{1}{2} =$

16) $5\frac{2}{3} + 1\frac{4}{7} =$

17) $4\frac{5}{6} + 6\frac{1}{4} =$

18) $2\frac{2}{5} + 3\frac{3}{8} =$

19) $3\frac{1}{6} + 2\frac{4}{9} =$

20) $5\frac{3}{5} + 3\frac{2}{3} =$

21) $4\frac{5}{8} + 1\frac{1}{3} =$

22) $6\frac{1}{9} + 4\frac{4}{5} =$

23) $2\frac{2}{7} + 3\frac{4}{5} =$

24) $3\frac{1}{2} + 1\frac{5}{7} =$

Subtracting Mix Numbers

Subtract the following fractions.

1) $7\frac{1}{3} - 6\frac{1}{3} =$

2) $4\frac{5}{8} - 4\frac{2}{8} =$

3) $8\frac{5}{9} - 7\frac{1}{9} =$

4) $4\frac{1}{4} - 1\frac{1}{3} =$

5) $3\frac{1}{3} - 2\frac{1}{6} =$

6) $8\frac{1}{2} - 3\frac{2}{5} =$

7) $7\frac{5}{8} - 3\frac{3}{8} =$

8) $9\frac{9}{13} - 4\frac{6}{13} =$

9) $5\frac{7}{12} - 2\frac{5}{12} =$

10) $4\frac{4}{7} - 1\frac{3}{7} =$

11) $7\frac{1}{5} - 2\frac{1}{10} =$

12) $4\frac{5}{6} - 2\frac{1}{6} =$

13) $6\frac{2}{45} - 1\frac{1}{5} =$

14) $4\frac{1}{2} - 2\frac{1}{4} =$

15) $14\frac{4}{5} - 11\frac{2}{5} =$

16) $6\frac{2}{4} - 1\frac{1}{4} =$

17) $4\frac{1}{5} - 2\frac{3}{5} =$

18) $5\frac{1}{8} - 2\frac{1}{2} =$

19) $6\frac{2}{3} - 1\frac{1}{9} =$

20) $4\frac{3}{5} - 4\frac{1}{15} =$

21) $9\frac{9}{11} - 5\frac{1}{2} =$

22) $8\frac{4}{5} - 2\frac{3}{20} =$

23) $3\frac{2}{3} - 2\frac{1}{9} =$

24) $7\frac{9}{14} - 3\frac{3}{14} =$

Simplify Fractions

Reduce these fractions to lowest terms

1) $\dfrac{15}{10} =$

2) $\dfrac{20}{30} =$

3) $\dfrac{28}{35} =$

4) $\dfrac{21}{28} =$

5) $\dfrac{6}{18} =$

6) $\dfrac{27}{63} =$

7) $\dfrac{16}{28} =$

8) $\dfrac{48}{60} =$

9) $\dfrac{8}{72} =$

10) $\dfrac{30}{12} =$

11) $\dfrac{45}{60} =$

12) $\dfrac{30}{90} =$

13) $\dfrac{18}{30} =$

14) $\dfrac{5}{20} =$

15) $\dfrac{16}{56} =$

16) $\dfrac{56}{84} =$

17) $\dfrac{88}{33} =$

18) $\dfrac{36}{135} =$

19) $\dfrac{21}{56} =$

20) $\dfrac{64}{56} =$

21) $\dfrac{140}{280} =$

22) $\dfrac{30}{155} =$

23) $\dfrac{210}{42} =$

24) $\dfrac{130}{520} =$

Multiplying Fractions

Find the product.

1) $\frac{4}{5} \times \frac{2}{6} =$

2) $\frac{4}{22} \times \frac{5}{8} =$

3) $\frac{8}{30} \times \frac{12}{16} =$

4) $\frac{9}{14} \times \frac{21}{36} =$

5) $\frac{14}{15} \times \frac{5}{7} =$

6) $\frac{16}{19} \times \frac{3}{4} =$

7) $\frac{4}{9} \times \frac{9}{8} =$

8) $\frac{47}{85} \times 0 =$

9) $\frac{5}{8} \times \frac{16}{6} =$

10) $\frac{28}{15} \times \frac{5}{7} =$

11) $\frac{32}{24} \times \frac{12}{16} =$

12) $\frac{6}{42} \times \frac{7}{36} =$

13) $\frac{13}{8} \times \frac{12}{4} =$

14) $\frac{10}{9} \times \frac{6}{5} =$

15) $\frac{35}{56} \times \frac{8}{7} =$

16) $\frac{16}{18} \times 9 =$

17) $\frac{5}{22} \times \frac{44}{15} =$

18) $\frac{10}{18} \times \frac{9}{20} =$

19) $\frac{7}{11} \times \frac{8}{21} =$

20) $\frac{26}{24} \times \frac{8}{52} =$

21) $\frac{6}{17} \times \frac{1}{12} =$

22) $\frac{20}{9} \times \frac{6}{100} =$

23) $\frac{8}{14} \times \frac{7}{72} =$

24) $\frac{50}{100} \times \frac{300}{400} =$

Multiplying Mixed Number

Multiply. Reduce to lowest terms.

1) $2\frac{3}{5} \times 1\frac{3}{4} =$

2) $1\frac{5}{6} \times 1\frac{1}{3} =$

3) $2\frac{3}{5} \times 1\frac{1}{7} =$

4) $3\frac{1}{7} \times 2\frac{1}{2} =$

5) $4\frac{3}{4} \times 1\frac{1}{4} =$

6) $3\frac{1}{2} \times 1\frac{4}{5} =$

7) $3\frac{3}{4} \times 1\frac{1}{2} =$

8) $5\frac{2}{3} \times 3\frac{1}{3} =$

9) $3\frac{2}{3} \times 3\frac{1}{2} =$

10) $2\frac{1}{3} \times 3\frac{1}{2} =$

11) $4\frac{3}{4} \times 3\frac{2}{3} =$

12) $3\frac{2}{4} \times 3\frac{1}{6} =$

13) $2\frac{2}{5} \times 1\frac{1}{3} =$

14) $2\frac{1}{3} \times 1\frac{1}{6} =$

15) $2\frac{2}{3} \times 3\frac{1}{2} =$

16) $2\frac{1}{8} \times 2\frac{2}{5} =$

17) $2\frac{1}{4} \times 1\frac{2}{3} =$

18) $2\frac{3}{5} \times 1\frac{1}{4} =$

19) $2\frac{3}{5} \times 1\frac{5}{6} =$

20) $3\frac{3}{5} \times 2\frac{3}{4} =$

21) $3\frac{3}{4} \times 1\frac{1}{3} =$

22) $2\frac{5}{8} \times 3\frac{1}{4} =$

Dividing Fractions

Divide these fractions.

1) $1 \div \frac{1}{5} =$

2) $\frac{7}{13} \div 7 =$

3) $\frac{5}{14} \div \frac{2}{5} =$

4) $\frac{15}{60} \div \frac{3}{4} =$

5) $\frac{2}{17} \div \frac{4}{17} =$

6) $\frac{4}{16} \div \frac{18}{24} =$

7) $0 \div \frac{1}{9} =$

8) $\frac{12}{16} \div \frac{8}{9} =$

9) $\frac{8}{12} \div \frac{4}{18} =$

10) $\frac{9}{14} \div \frac{3}{7} =$

11) $\frac{8}{15} \div \frac{25}{16} =$

12) $\frac{35}{12} \div \frac{15}{6} =$

13) $\frac{9}{15} \div \frac{9}{5} =$

14) $\frac{8}{18} \div \frac{40}{6} =$

15) $\frac{45}{21} \div \frac{9}{21} =$

16) $\frac{7}{30} \div \frac{63}{5} =$

17) $\frac{36}{8} \div \frac{18}{24} =$

18) $9 \div \frac{1}{2} =$

19) $\frac{48}{35} \div \frac{8}{7} =$

20) $\frac{3}{36} \div \frac{9}{6} =$

21) $\frac{4}{7} \div \frac{12}{14} =$

22) $\frac{8}{40} \div \frac{10}{5} =$

Dividing Mixed Number

Divide the following mixed numbers. Cancel and simplify when possible.

1) $4\frac{1}{4} \div 4\frac{1}{3} =$

2) $2\frac{1}{6} \div 1\frac{2}{2} =$

3) $5\frac{1}{3} \div 3\frac{3}{4} =$

4) $3\frac{1}{6} \div 3\frac{1}{5} =$

5) $5\frac{1}{6} \div 1\frac{2}{3} =$

6) $3\frac{3}{5} \div 2\frac{2}{6} =$

7) $4\frac{3}{5} \div 2\frac{1}{3} =$

8) $2\frac{4}{9} \div 1\frac{1}{9} =$

9) $3\frac{5}{6} \div 3\frac{1}{2} =$

10) $9\frac{1}{9} \div 3\frac{2}{3} =$

11) $2\frac{2}{7} \div 4\frac{1}{7} =$

12) $4\frac{3}{8} \div 1\frac{3}{4} =$

13) $5\frac{1}{8} \div 1\frac{1}{12} =$

14) $6\frac{3}{8} \div 3\frac{1}{3} =$

15) $4\frac{2}{5} \div 1\frac{1}{5} =$

16) $2\frac{1}{2} \div 2\frac{2}{9} =$

17) $7\frac{1}{6} \div 5\frac{3}{8} =$

18) $5\frac{1}{2} \div 4\frac{1}{3} =$

19) $4\frac{5}{7} \div 1\frac{1}{3} =$

20) $3\frac{5}{6} \div 1\frac{1}{4} =$

21) $9\frac{1}{2} \div 7\frac{1}{3} =$

22) $3\frac{1}{8} \div 1\frac{1}{9} =$

23) $4\frac{1}{4} \div 3\frac{3}{4} =$

24) $3\frac{1}{6} \div 3\frac{1}{3} =$

Comparing Fractions

Compare the fractions, and write >, < or =

1) $\dfrac{14}{3}$ ____ $\dfrac{24}{15}$

2) $\dfrac{32}{3}$ ____ $\dfrac{2}{5}$

3) $\dfrac{4}{9}$ ____ $\dfrac{2}{4}$

4) $\dfrac{12}{4}$ ____ $\dfrac{13}{9}$

5) $\dfrac{1}{8}$ ____ $\dfrac{2}{3}$

6) $\dfrac{10}{6}$ ____ $\dfrac{16}{7}$

7) $\dfrac{12}{13}$ ____ $\dfrac{7}{9}$

8) $\dfrac{20}{14}$ ____ $\dfrac{25}{3}$

9) $4\dfrac{1}{12}$ ____ $6\dfrac{1}{3}$

10) $8\dfrac{1}{6}$ ____ $3\dfrac{1}{8}$

11) $3\dfrac{1}{2}$ ____ $3\dfrac{1}{5}$

12) $7\dfrac{5}{8}$ ____ $7\dfrac{2}{9}$

13) $3\dfrac{2}{8}$ ____ $5\dfrac{3}{5}$

14) $\dfrac{1}{15}$ ____ $\dfrac{3}{7}$

15) $\dfrac{31}{25}$ ____ $\dfrac{19}{83}$

16) $\dfrac{12}{100}$ ____ $\dfrac{6}{62}$

17) $15\dfrac{1}{4}$ ____ $15\dfrac{1}{9}$

18) $\dfrac{1}{5}$ ____ $\dfrac{1}{9}$

19) $\dfrac{1}{7}$ ____ $\dfrac{1}{13}$

20) $\dfrac{1}{18}$ ____ $\dfrac{8}{15}$

21) $\dfrac{7}{22}$ ____ $\dfrac{9}{76}$

22) $\dfrac{4}{5}$ ____ $\dfrac{2}{5}$

23) $2\dfrac{17}{14}$ ____ $3\dfrac{3}{14}$

24) $3\dfrac{25}{4}$ ____ $4\dfrac{5}{4}$

Answer key Chapter 4

Adding Fractions – Like Denominator

1) $\frac{3}{4}$
2) $\frac{3}{5}$
3) $\frac{3}{8}$
4) $\frac{5}{11}$
5) $\frac{5}{21}$
6) $\frac{11}{49}$
7) $\frac{13}{7}$
8) $\frac{4}{15}$
9) $\frac{9}{19}$
10) $\frac{2}{13}$
11) $\frac{2}{5}$
12) $\frac{10}{17}$
13) $\frac{19}{20}$
14) $\frac{11}{25}$
15) $\frac{9}{14}$
16) $\frac{17}{30}$
17) $\frac{2}{9}$
18) $\frac{32}{5}$
19) $\frac{23}{6}$
20) $\frac{36}{37}$

Adding Fractions – Unlike Denominator

1) $\frac{11}{12}$
2) $\frac{13}{10}$
3) $\frac{27}{28}$
4) $\frac{27}{22}$
5) $\frac{13}{18}$
6) $\frac{14}{27}$
7) $\frac{19}{24}$
8) $\frac{11}{20}$
9) $\frac{21}{22}$
10) $\frac{43}{63}$
11) $\frac{47}{72}$
12) $\frac{31}{32}$
13) $\frac{7}{13}$
14) $\frac{5}{9}$
15) $\frac{27}{64}$
16) $\frac{2}{3}$
17) $\frac{15}{14}$
18) $\frac{7}{24}$
19) $\frac{13}{75}$
20) $\frac{19}{24}$
21) $\frac{9}{44}$
22) $\frac{59}{60}$
23) $\frac{7}{24}$
24) $\frac{7}{54}$

Subtracting Fractions – Like Denominator

1) 1
2) $\frac{1}{4}$
3) $\frac{3}{14}$
4) 2
5) $\frac{2}{17}$
6) $\frac{8}{33}$
7) $\frac{6}{25}$
8) $\frac{5}{9}$
9) $\frac{2}{5}$
10) $\frac{4}{7}$
11) $\frac{8}{5}$
12) $\frac{5}{19}$
13) $\frac{2}{3}$
14) $\frac{11}{43}$
15) $\frac{1}{7}$
16) $\frac{3}{29}$
17) $\frac{1}{5}$
18) $\frac{4}{53}$

Common Core Math Workbook

19) $\frac{5}{31}$
20) $\frac{2}{39}$
21) $\frac{2}{13}$
22) $\frac{2}{23}$
23) $\frac{1}{8}$
24) $\frac{12}{65}$

Subtracting Fractions – Unlike Denominator

1) $\frac{11}{20}$
2) $\frac{5}{12}$
3) $\frac{17}{42}$
4) $\frac{1}{4}$
5) $\frac{9}{14}$
6) $\frac{7}{36}$
7) $\frac{9}{20}$
8) $\frac{29}{48}$
9) $\frac{26}{63}$
10) $\frac{7}{32}$
11) $\frac{3}{5}$
12) $\frac{3}{5}$
13) $\frac{2}{3}$
14) $\frac{17}{15}$
15) $\frac{1}{22}$
16) $\frac{41}{54}$
17) $\frac{7}{96}$
18) $\frac{3}{20}$
19) $\frac{5}{18}$
20) $\frac{4}{33}$

Converting Mix Numbers

1) $\frac{23}{6}$
2) $\frac{86}{15}$
3) $\frac{13}{3}$
4) $\frac{18}{7}$
5) $\frac{29}{4}$
6) $\frac{82}{21}$
7) $\frac{59}{10}$
8) $\frac{55}{12}$
9) $\frac{43}{11}$
10) $\frac{32}{5}$
11) $\frac{26}{3}$
12) $\frac{35}{12}$
13) $\frac{23}{6}$
14) $\frac{52}{11}$
15) $\frac{29}{4}$
16) $\frac{61}{11}$
17) $\frac{41}{5}$
18) $\frac{43}{12}$
19) $\frac{133}{22}$
20) $\frac{11}{3}$
21) $\frac{39}{5}$
22) $\frac{39}{8}$
23) $\frac{41}{6}$
24) $\frac{129}{10}$

Converting improper Fractions

1) $4\frac{6}{14}$
2) $2\frac{24}{37}$
3) $2\frac{15}{17}$
4) $2\frac{11}{23}$
5) $4\frac{7}{16}$
6) $3\frac{11}{42}$
7) $3\frac{21}{33}$
8) $5\frac{1}{5}$
9) $1\frac{14}{19}$
10) $6\frac{1}{2}$
11) $9\frac{3}{4}$
12) $3\frac{15}{65}$
13) $1\frac{12}{64}$
14) $2\frac{4}{7}$
15) $8\frac{6}{13}$

16) $12\frac{1}{4}$

17) $13\frac{5}{9}$

18) $5\frac{1}{12}$

19) $6\frac{1}{6}$

20) $3\frac{1}{9}$

21) $1\frac{1}{4}$

22) $6\frac{1}{13}$

23) $5\frac{1}{8}$

24) $9\frac{1}{7}$

Adding Mix Numbers

1) $5\frac{3}{5}$

2) 9

3) $5\frac{3}{8}$

4) $9\frac{1}{8}$

5) $5\frac{25}{28}$

6) $9\frac{9}{10}$

7) $4\frac{11}{27}$

8) $6\frac{1}{12}$

9) 6

10) $5\frac{1}{7}$

11) $6\frac{9}{10}$

12) $8\frac{1}{12}$

13) 8

14) 7

15) $9\frac{1}{10}$

16) $7\frac{5}{21}$

17) $11\frac{1}{12}$

18) $5\frac{31}{40}$

19) $5\frac{11}{18}$

20) $9\frac{4}{15}$

21) $5\frac{23}{24}$

22) $10\frac{41}{45}$

23) $6\frac{3}{35}$

24) $5\frac{3}{14}$

Subtracting Mix Numbers

1) 1

2) $\frac{3}{8}$

3) $1\frac{4}{9}$

4) $2\frac{11}{12}$

5) $1\frac{1}{6}$

6) $5\frac{1}{10}$

7) $4\frac{1}{4}$

8) $5\frac{3}{13}$

9) $3\frac{1}{6}$

10) $3\frac{1}{7}$

11) $5\frac{1}{10}$

12) $2\frac{2}{3}$

13) $4\frac{38}{45}$

14) $2\frac{1}{4}$

15) $3\frac{2}{5}$

16) $5\frac{1}{4}$

17) $1\frac{3}{5}$

18) $2\frac{5}{8}$

19) $5\frac{5}{9}$

20) $\frac{8}{15}$

21) $4\frac{7}{22}$

22) $6\frac{13}{20}$

23) $1\frac{5}{9}$

24) $4\frac{3}{7}$

Simplify Fractions

1) $\frac{3}{2}$

2) $\frac{2}{3}$

3) $\frac{4}{5}$

4) $\frac{3}{4}$

5) $\frac{1}{3}$

6) $\frac{3}{7}$

7) $\frac{4}{7}$

8) $\frac{4}{5}$

9) $\frac{1}{9}$

10) $\frac{5}{2}$

11) $\frac{3}{4}$

12) $\frac{1}{3}$

13) $\frac{3}{5}$

14) $\frac{1}{4}$

15) $\frac{2}{7}$

16) $\frac{2}{3}$

17) $\frac{8}{3}$

18) $\frac{4}{15}$

19) $\frac{3}{8}$

20) $\frac{8}{7}$

21) $\frac{1}{2}$

22) $\frac{6}{31}$

23) 5

24) $\frac{1}{4}$

Multiplying Fractions

1) $\frac{4}{15}$

2) $\frac{5}{44}$

3) $\frac{1}{5}$

4) $\frac{3}{8}$

5) $\frac{2}{3}$

6) $\frac{12}{19}$

7) $\frac{1}{2}$

8) 0

9) $\frac{5}{3}$

10) $\frac{4}{3}$

11) 1

12) $\frac{1}{36}$

13) $\frac{39}{8}$

14) $\frac{4}{3}$

15) $\frac{5}{7}$

16) 8

17) $\frac{2}{3}$

18) $\frac{1}{4}$

19) $\frac{8}{33}$

20) $\frac{1}{6}$

21) $\frac{1}{34}$

22) $\frac{2}{15}$

23) $\frac{1}{18}$

24) $\frac{3}{8}$

Multiplying Mixed Number

1) $4\frac{11}{20}$

2) $2\frac{4}{9}$

3) $2\frac{34}{35}$

4) $7\frac{6}{7}$

5) $5\frac{15}{16}$

6) $6\frac{3}{10}$

7) $5\frac{5}{8}$

8) $18\frac{8}{9}$

9) $12\frac{5}{6}$

10) $8\frac{1}{6}$

11) $17\frac{5}{12}$

12) $11\frac{1}{12}$

13) $3\frac{1}{5}$

14) $2\frac{13}{18}$

15) $9\frac{1}{3}$

16) $5\frac{1}{10}$

17) $3\frac{3}{4}$

18) $3\frac{1}{4}$

19) $4\frac{23}{30}$

20) $9\frac{9}{10}$

21) 5

22) $8\frac{17}{32}$

Dividing Fractions

1) 5
2) $\frac{1}{13}$
3) $\frac{25}{28}$
4) $\frac{1}{3}$
5) $\frac{1}{2}$
6) $\frac{1}{3}$
7) 0
8) $\frac{27}{32}$
9) 3
10) $\frac{3}{2}$
11) $\frac{128}{375}$
12) $\frac{7}{6}$
13) $\frac{1}{3}$
14) $\frac{1}{15}$
15) 5
16) $\frac{1}{54}$
17) 6
18) 18
19) $\frac{6}{5}$
20) $\frac{1}{18}$
21) $\frac{2}{3}$
22) $\frac{1}{10}$

Dividing Mixed Number

1) $\frac{51}{52}$
2) $\frac{1}{12}$
3) $1\frac{19}{45}$
4) $\frac{95}{96}$
5) $3\frac{1}{10}$
6) $1\frac{19}{35}$
7) $1\frac{34}{35}$
8) $\frac{2}{81}$
9) $1\frac{2}{21}$
10) $2\frac{16}{33}$
11) $\frac{16}{29}$
12) $2\frac{1}{2}$
13) $4\frac{19}{26}$
14) $1\frac{73}{80}$
15) $3\frac{2}{3}$
16) $1\frac{1}{8}$
17) $1\frac{1}{3}$
18) $1\frac{7}{26}$
19) $3\frac{15}{28}$
20) $3\frac{1}{15}$
21) $1\frac{13}{44}$
22) $2\frac{13}{16}$
23) $1\frac{2}{15}$
24) $\frac{19}{20}$

Comparing Fractions

1) >
2) >
3) <
4) >
5) <
6) <
7) >
8) <
9) <
10) >
11) >
12) >
13) <
14) <
15) >
16) <
17) >
18) >
19) >
20) <
21) >
22) >
23) =
24) >

Chapter 5:

Decimal

Graph Decimals

Write the decimals indicated by the arrows.

1)

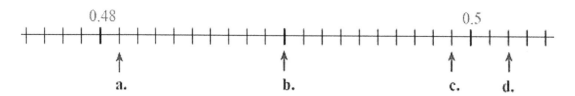

a. _____ b. _____ c. _____ d. _____

2)

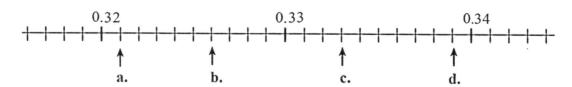

a. _____ b. _____ c. _____ d. _____

3)

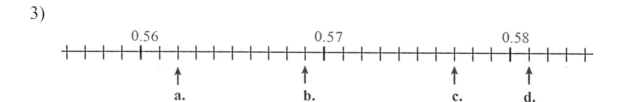

a. _____ b. _____ c. _____ d. _____

4)

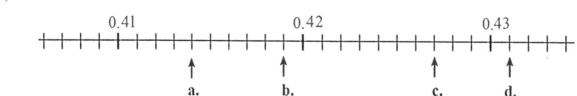

a. _____ b. _____ c. _____ d. _____

Round Decimals

Round each number to the correct place value

1) 0.8<u>3</u> =

2) 3.<u>0</u>2 =

3) 7.<u>7</u>11 =

4) 0.<u>4</u>78 =

5) <u>8</u>.824 =

6) 0.0<u>7</u>8 =

7) 8.<u>1</u>3 =

8) 84.8<u>4</u>0 =

9) 2.5<u>3</u>8 =

10) 12.<u>2</u>97 =

11) 2.<u>0</u>8 =

12) 5.<u>3</u>24 =

13) 2.<u>1</u>32 =

14) 8.0<u>7</u>32 =

15) 5<u>5</u>.78 =

16) 2<u>8</u>.24 =

17) 5<u>2</u>7.156 =

18) 624.<u>7</u>88 =

19) 17.4<u>8</u>1 =

20) 9<u>4</u>.86 =

21) 4.3<u>0</u>67 =

22) 57.<u>0</u>86 =

23) 224.<u>2</u>24 =

24) 0.1<u>3</u>44 =

25) 0.00<u>6</u>9 =

26) 9.0<u>3</u>86 =

27) 35.5<u>4</u>22 =

28) 11.0<u>9</u>31 =

Decimals Addition

Add the following.

1) 32.12 + 24.28

2) 0.88 + 0.21

3) 15.36 + 10.87

4) 75.165 + 4.105

5) 8.650 + 7.82

6) 5.324 + 2.138

7) 81.21 + 15.85

8) 56.25 + 22.35

9) 46.21 + 10.07

10) 8.96 + 11.23

11) 15.214 + 11.251

12) 72.36 + 5.32

13) 32.05 + 8.54

14) 137.21 + 2.75

Decimals Subtraction

Subtract the following

1) 9.35
   -3.52
   _____

2) 75.35
   -62.37
   _____

3) 0.68
   -0.4
   _____

4) 11.245
   -8.6
   _____

5) 0.652
   -0.09
   _____

6) 75.25
   -28.88
   _____

7) 112.66
   -88.98
   _____

8) 32.56
   -12.45
   _____

9) 68.35
   -59.98
   _____

10) 6.985
    -0.223
    _____

11) 55.69
    -45.32
    _____

12) 12.352
    -2.325
    _____

13) 19.231
    -4.128
    _____

14) 128.98
    -7.92
    _____

Decimals Multiplication

Solve.

1) 3.1 × 3.4

2) 7.5 × 4.5

3) 5.04 × 3.04

4) 88.09 × 100

5) 23.9 × 10

6) 35.62 × 5.5

7) 32.75 × 11.3

8) 2.65 × 8.35

9) 12.05 × 0.04

10) 24.04 × 8.08

11) 12.34 × 11.2

12) 6.37 × 0.02

13) 9.4 × 0.14

14) 15.4 × 6.05

Decimal Division

Dividing Decimals.

1) $8 \div 10{,}000 =$

2) $4 \div 100 =$

3) $3.4 \div 100 =$

4) $0.002 \div 10 =$

5) $8 \div 64 =$

6) $3 \div 81 =$

7) $5 \div 45 =$

8) $9 \div 180 =$

9) $7 \div 1{,}000 =$

10) $0.6 \div 0.63 =$

11) $0.9 \div 0.009 =$

12) $0.6 \div 0.12 =$

13) $0.6 \div 0.42 =$

14) $0.4 \div 0.04 =$

15) $3.08 \div 10 =$

16) $9.4 \div 10 =$

17) $6.75 \div 100 =$

18) $18.3 \div 3.3 =$

19) $64.4 \div 4 =$

20) $0.4 \div 0.004 =$

21) $7.05 \div 3.5 =$

22) $0.08 \div 0.40 =$

23) $0.9 \div 7.6 =$

24) $0.09 \div 54 =$

25) $5.24 \div 0.5 =$

26) $0.025 \div 125 =$

Comparing Decimals

Write the Correct Comparison Symbol (>, < or =)

1) 1.42 ____ 2.42

2) 0.5 ____ 0.425

3) 13.6 ____ 13.600

4) 7.07 ____ 7.70

5) 0.922 ____ 0.92

6) 0.856 ____ 0.956

7) 4.34 ____ 4.242

8) 5.0025 ____ 5.025

9) 24.087 ____ 24.078

10) 7.12 ____ 7.29

11) 4.44 ____ 4.444

12) 0.09 ____ 0.18

13) 1.302 ____ 1.32

14) 9.56 ____ 9.0569

15) 0.33 ____ 0.033

16) 21.04 ____ 21.040

17) 0.250 ____ 0.35

18) 44.92 ____ 45.01

19) 0.085 ____ 0.805

20) 36.5 ____ 29.8

21) 7.89 ____ 10.2

22) 0.024 ____ 0.0204

23) 5.042 ____ 0.5042

24) 7.5 ____ 0.758

25) 6.5 ____ 0.659

26) 3.24 ____ 3.2400

27) 8.34 ____ 0.834

28) 2.0809 ____ 2.0890

Convert Fraction to Decimal

Write each as a decimal.

1) $\frac{50}{100} =$

2) $\frac{46}{100} =$

3) $\frac{8}{50} =$

4) $\frac{8}{32} =$

5) $\frac{8}{72} =$

6) $\frac{56}{100} =$

7) $\frac{4}{50} =$

8) $\frac{31}{48} =$

9) $\frac{27}{300} =$

10) $\frac{15}{55} =$

11) $\frac{16}{32} =$

12) $\frac{6}{16} =$

13) $\frac{3}{10} =$

14) $\frac{18}{250} =$

15) $\frac{24}{80} =$

16) $\frac{30}{40} =$

17) $\frac{68}{100} =$

18) $\frac{7}{35} =$

19) $\frac{87}{100} =$

20) $\frac{1}{100} =$

21) $\frac{6}{36} =$

22) $\frac{2}{80} =$

Answer key Chapter 5

Graph Decimals

1) a. 0.481   b. 0.49   c. 0.499   d. 0.502
2) a. 0.321   b. 0.326   c. 0.333   d. 0339
3) a. 0.562   b. 0.569   c. 0.577   d. 0.581
4) a. 0.414   b. 0.419   c. 0.427   d. 0431

Round Decimals

1) 0.8
2) 3.0
3) 7.7
4) 0.5
5) 9.0
6) 0.08
7) 8.1
8) 84.84
9) 2.54
10) 12.3
11) 2.1
12) 5.3
13) 2.1
14) 8.07
15) 56.0
16) 28.0
17) 530.0
18) 624.8
19) 17.48
20) 95.0
21) 4.31
22) 57.1
23) 224.2
24) 0.13
25) 0.007
26) 9.04
27) 35.54
28) 11.09

Decimals Addition

1) 56.4
2) 1.09
3) 26.23
4) 79.27
5) 16.47
6) 7.462
7) 97.06
8) 78.6
9) 56.28
10) 20.19
11) 26.465
12) 77.68
13) 40.59
14) 139.96

Decimals Subtraction

1) 5.83
2) 12.98
3) 0.28
4) 2.645
5) 0.562
6) 46.37
7) 23.68
8) 20.11
9) 8.37
10) 6.762
11) 10.37
12) 10.027
13) 15.103
14) 121.06

Decimals Multiplication

1) 10.54
2) 33.75
3) 15.3216

Common Core Math Workbook

4) 8,809
5) 239
6) 195.91
7) 370.075
8) 22.1275
9) 0.482
10) 194.2432
11) 138.208
12) 0.1274
13) 1.316
14) 93.17

Decimal Division

1) 0.0008
2) 0.04
3) 0.034
4) 0.0002
5) 0.125
6) 0.037....
7) 0.111…
8) 0.05
9) 0.007
10) 0.952…
11) 100
12) 5
13) 1.4285…
14) 10
15) 0.308
16) 0.94
17) 0.0675
18) 5.5454…
19) 16.1
20) 100
21) 2.01428…
22) 0.2
23) 0.1184…
24) 0.0016
25) 10.48
26) 0.0002

Comparing Decimals

1) <
2) >
3) =
4) <
5) >
6) <
7) >
8) <
9) >
10) >
11) <
12) <
13) <
14) >
15) >
16) =
17) <
18) <
19) <
20) >
21) <
22) >
23) >
24) >
25) >
26) =
27) >
28) <

Convert Fraction to Decimal

1) 0.5
2) 0.46
3) 0.16
4) 0.25
5) 0.11
6) 0.56
7) 0.08
8) 0.646
9) 0.09
10) 0.27
11) 0.5
12) 0.375
13) 0.3
14) 0.072
15) 0.3

16) 0.75

17) 0.68

18) 0.2

19) 0.87

20) 0.01

21) 0.166

22) 0.025

Chapter 6: Exponent and Radicals

Common Core Math Workbook

Positive Exponents

Simplify. Your answer should contain only positive exponents.

1) $3^4 =$

2) $2^5 =$

3) $\frac{3x^6y}{xy} =$

4) $(12x^2x)^3 =$

5) $(x^2)^4 =$

6) $(\frac{1}{4})^3 =$

7) $0^8 =$

8) $6 \times 6 \times 6 =$

9) $3 \times 3 \times 3 \times 3 \times 3 =$

10) $(4x^4y)^2 =$

11) $9^3 =$

12) $(5x^3y^2)^2 =$

13) $5 \times 10^4 =$

14) $0.6 \times 0.6 \times 0.6 =$

15) $\frac{1}{3} \times \frac{1}{3} \times \frac{1}{3} =$

16) $4^5 =$

17) $(5x^8y^2)^3 =$

18) $7^3 =$

19) $y \times y \times y \times y =$

20) $8 \times 8 \times 8 \times 8 =$

21) $(2x^4y^2z)^3 =$

22) $8^0 =$

23) $(11x^4y^{-1})^4 =$

24) $(2x^2y^4)^5 =$

Negative Exponents

Simplify. Leave no negative exponents.

1) $2^{-4} =$

2) $9^{-2} =$

3) $(\frac{1}{3})^{-2} =$

4) $8^{-3} =$

5) $1^{150} =$

6) $6^{-3} =$

7) $(\frac{1}{2})^{-6} =$

8) $-8y^{-4} =$

9) $(\frac{1}{y^{-5}})^{-3} =$

10) $x^{-\frac{4}{5}} =$

11) $\frac{1}{7^{-6}} =$

12) $3^{-5} =$

13) $5^{-2} =$

14) $13^{-2} =$

15) $30^{-2} =$

16) $x^{-8} =$

17) $(x^2)^{-4} =$

18) $x^{-2} \times x^{-2} \times x^{-2} \times x^{-2} =$

19) $\frac{1}{3} \times \frac{1}{3} =$

20) $100^{-2} =$

21) $100z^{-3} =$

22) $3^{-4} =$

23) $(-\frac{1}{11})^2 =$

24) $14^0 =$

25) $(\frac{1}{x})^{-18} =$

26) $15^{-2} =$

Add and subtract Exponents

Solve each problem.

1) $4^2 + 5^3 =$

2) $x^8 + x^8 =$

3) $5b^3 - 4b^3 =$

4) $6 + 5^2 =$

5) $9 - 6^2 =$

6) $12 + 3^2 =$

7) $5x^2 + 8x^2 =$

8) $9^2 + 2^6 =$

9) $3^6 - 4^3 =$

10) $8^2 - 10^0 =$

11) $7^2 - 4^2 =$

12) $9^2 + 3^4 =$

13) $12^2 - 5^2 =$

14) $7^2 + 7^2 =$

15) $6^3 - 4^3 =$

16) $1^{24} + 1^{28} =$

17) $4^3 - 2^3 =$

18) $5^4 - 5^2 =$

19) $7^2 - 4^2 =$

20) $5^2 + 8^2 =$

21) $4^2 + 3^4 =$

22) $18 + 2^4 =$

23) $7x^8 + 5x^8 =$

24) $9^0 + 8^2 =$

25) $5^2 + 5^2 =$

26) $10^3 + 2^2 =$

27) $(\frac{1}{3})^2 + (\frac{1}{3})^2 =$

28) $8^2 + 2^2 =$

Exponent multiplication

Simplify each of the following

1) $2^5 \times 2^3 =$

2) $7^2 \times 8^0 =$

3) $9^1 \times 4^2 =$

4) $a^{-5} \times a^{-5} =$

5) $y^{-3} \times y^{-3} \times y^{-3} =$

6) $4^5 \times 5^7 \times 4^{-4} \times 5^{-6} =$

7) $6x^4y^3 \times 4x^3y^2 =$

8) $(x^3)^5 =$

9) $(x^4y^6)^5 \times (x^4y^5)^{-5} =$

10) $8^4 \times 8^2 =$

11) $a^{4b} \times a^0 =$

12) $4^2 \times 4^2 =$

13) $a^{3m} \times a^{2n} =$

14) $2a^n \times 4b^n =$

15) $5^{-3} \times 4^{-3} =$

16) $6^{10} \times 3^{10} =$

17) $(7^6)^5 =$

18) $\left(\frac{1}{6}\right)^2 \times \left(\frac{1}{6}\right)^4 \times \left(\frac{1}{6}\right)^5 =$

19) $\left(\frac{1}{9}\right)^{52} \times 9^{52} =$

20) $(4m)^{\frac{4}{5}} \times (-2m)^{\frac{4}{5}} =$

21) $(x^4y)^{\frac{1}{4}} \times (xy^3)^{\frac{1}{4}} =$

22) $(2a^m b^n)^r =$

23) $(5x^3y^2)^3 =$

24) $(x^{\frac{1}{3}}y^2)^{\frac{-1}{3}} \times (x^4y^6)^0 =$

25) $7^6 \times 7^5 =$

26) $28^{\frac{1}{6}} \times 28^{\frac{1}{3}} =$

27) $9^5 \times 3^5 =$

28) $(x^{12})^0 =$

Exponent division

Simplify. Your answer should contain only positive exponents.

1) $\dfrac{5^4}{5} =$

2) $\dfrac{38x^4}{x} =$

3) $\dfrac{a^m}{a^{2n}} =$

4) $\dfrac{3x^{-6}}{15x^{-4}} =$

5) $\dfrac{63x^9}{7x^4} =$

6) $\dfrac{17x^7}{5x^8} =$

7) $\dfrac{36x^8}{12y^3} =$

8) $\dfrac{45xy^6}{x^4y^2} =$

9) $\dfrac{3x^9}{8x} =$

10) $\dfrac{45x^7y^9}{5x^8} =$

11) $\dfrac{12x^4}{20x^9y^{12}} =$

12) $\dfrac{8yx^7}{40yx^{10}} =$

13) $\dfrac{21x^3y^2}{3x^2y^3} =$

14) $\dfrac{x^{4.75}}{x^{0.75}} =$

15) $\dfrac{9x^4y}{18xy^3} =$

16) $\dfrac{34b^3r^8}{17a^2b^5} =$

17) $\dfrac{30x^7}{15x^9} =$

18) $\dfrac{44x^5}{11x^8} =$

19) $\dfrac{6^5}{6^3} =$

20) $\dfrac{x}{x^{10}} =$

21) $\dfrac{13^7}{13^4} =$

22) $\dfrac{3xy^5}{12y^3} =$

23) $\dfrac{13x^6y}{169xy^3} =$

24) $\dfrac{48x^5}{8y^9} =$

Scientific Notation

Write each number in scientific notation.

1) $9,500,000 =$

2) $800 =$

3) $0.000007 =$

4) $387,000 =$

5) $0.00139 =$

6) $0.85 =$

7) $0.000093 =$

8) $20,000,000 =$

9) $28,000,000 =$

10) $230,000,000 =$

11) $0.000049 =$

12) $0.00002 =$

13) $0.00027 =$

14) $70,000 =$

15) $2,870 =$

16) $190,000 =$

17) $0.0223 =$

18) $0.7 =$

19) $0.082 =$

20) $310,000 =$

21) $48,000 =$

22) $0.000098 =$

23) $0.035 =$

24) $1,778 =$

25) $58,781 =$

26) $24,500 =$

27) $33,021 =$

28) $8,100,000 =$

Square Roots

Find the square root of each number.

1) $\sqrt{64} =$

2) $\sqrt{0} =$

3) $\sqrt{324} =$

4) $\sqrt{484} =$

5) $\sqrt{1,600} =$

6) $\sqrt{529} =$

7) $\sqrt{0.01} =$

8) $\sqrt{10,000} =$

9) $\sqrt{0.16} =$

10) $\sqrt{0.36} =$

11) $\sqrt{0.25} =$

12) $\sqrt{1.21} =$

13) $\sqrt{784} =$

14) $\sqrt{576} =$

15) $\sqrt{676} =$

16) $\sqrt{961} =$

17) $\sqrt{1,681} =$

18) $\sqrt{0.81} =$

19) $\sqrt{0.49} =$

20) $\sqrt{0.64} =$

21) $\sqrt{1,089} =$

22) $\sqrt{2,500} =$

23) $\sqrt{8,100} =$

24) $\sqrt{12,100} =$

25) $\sqrt{2.25} =$

26) $\sqrt{1.69} =$

27) $\sqrt{1.44} =$

28) $\sqrt{0.04} =$

Simplify Square Roots

Simplify the following.

1) $\sqrt{54} =$

2) $\sqrt{108} =$

3) $\sqrt{12} =$

4) $\sqrt{99} =$

5) $\sqrt{200} =$

6) $\sqrt{45} =$

7) $8\sqrt{50} =$

8) $3\sqrt{300} =$

9) $\sqrt{24} =$

10) $2\sqrt{18} =$

11) $4\sqrt{3} + 7\sqrt{3} =$

12) $\frac{11}{4+\sqrt{5}} =$

13) $\sqrt{48} =$

14) $\frac{4}{3-\sqrt{5}} =$

15) $\sqrt{18} \times \sqrt{2} =$

16) $\frac{\sqrt{300}}{\sqrt{3}} =$

17) $\frac{\sqrt{90}}{\sqrt{18 \times 5}} =$

18) $\sqrt{80y^6} =$

19) $6\sqrt{81a} =$

20) $\sqrt{41+8} + \sqrt{9} =$

21) $\sqrt{72} =$

22) $\sqrt{432} =$

23) $\sqrt{112} =$

24) $\sqrt{128} =$

25) $\sqrt{768} =$

26) $\sqrt{96} =$

Answer key Chapter 6

Positive Exponents

1) 81
2) 32
3) $3x^5$
4) $1,728x^9$
5) x^8
6) $\frac{1}{64}$
7) 0
8) 6^3
9) 3^5
10) $16x^8y^2$
11) 729
12) $25x^6y^4$
13) 50,000
14) 0.6^3
15) $(\frac{1}{3})^3$
16) 1,024
17) $125x^{24}y^6$
18) 343
19) y^4
20) 8^4
21) $8x^{12}y^6z^3$
22) 1
23) $\frac{121x^8}{y^4}$
24) $32x^{10}y^{20}$

Negative Exponents

1) $\frac{1}{16}$
2) $\frac{1}{81}$
3) 9
4) $\frac{1}{512}$
5) 1
6) $\frac{1}{216}$
7) 64
8) $\frac{-8}{y^4}$
9) y^{15}
10) $\frac{1}{x^{\frac{4}{5}}}$
11) 7^6
12) $\frac{1}{243}$
13) $\frac{1}{25}$
14) $\frac{1}{169}$
15) $\frac{1}{900}$
16) $\frac{1}{x^8}$
17) $\frac{1}{x^8}$
18) $\frac{1}{x^8}$
19) $\frac{1}{3^2}$
20) $\frac{1}{10,000}$
21) $\frac{100}{z^3}$
22) $\frac{1}{81}$
23) $\frac{1}{121}$
24) 1
25) x^{18}
26) $\frac{1}{225}$

Add and subtract Exponents

1) 141
2) $2x^8$
3) b^3
4) 31
5) -27
6) 21
7) $13x^2$
8) 145
9) 665
10) 63
11) 33
12) 162
13) 119
14) 98
15) 152

16) 1
17) 56
18) 600
19) 33
20) 89
21) 97
22) 34
23) $12x^8$
24) 65
25) 50
26) 1,004
27) $\frac{2}{9}$
28) 68

Exponent multiplication

1) 2^8
2) 49
3) 144
4) a^{-10}
5) y^{-9}
6) 20
7) $24x^7y^5$
8) x^{15}
9) y^5
10) 8^6
11) a^{4b}
12) 4^4
13) a^{3m+2n}
14) $8(ab)^n$
15) 20^{-3}
16) 18^{10}
17) 7^{30}
18) $(\frac{1}{6})^{11}$
19) 1
20) $(-8m^2)^{\frac{4}{5}}$
21) $x^{\frac{5}{4}}y$
22) $2^r a^{mr} b^{nr}$
23) $125x^9y^6$
24) $x^{\frac{5}{4}}y$
25) 7^{11}
26) $28^{\frac{1}{2}}$
27) $27^5 = 3^{15}$
28) 1

Exponent division

1) 5^3
2) $38x^3$
3) a^{m-2n}
4) $\frac{1}{5x^2}$
5) $9x^5$
6) $\frac{17}{5x}$
7) $\frac{3x^8}{y^3}$
8) $\frac{45y^4}{x^3}$
9) $\frac{3x^8}{8}$
10) $\frac{9y^9}{x}$
11) $\frac{3}{5x^5y^{12}}$
12) $\frac{1}{5x^3}$
13) $\frac{7x}{y}$
14) x^4
15) $\frac{x^3}{2y^2}$
16) $\frac{2r^8}{a^2b^2}$
17) $\frac{2}{x^2}$
18) $\frac{4}{x^3}$
19) 6^2
20) $\frac{1}{x^9}$
21) 13^3
22) $\frac{1}{4}xy^2$
23) $\frac{x^5}{13y^2}$
24) $\frac{6x^5}{y^9}$

Scientific Notation

1) 9.5×10^6
2) 8×10^2
3) 7×10^{-6}
4) 3.87×10^5
5) 1.39×10^{-3}
6) 8.5×10^{-1}

Common Core Math Workbook

7) 9.3×10^{-5}
8) 2×10^7
9) 2.8×10^7
10) 2.3×10^8
11) 4.9×10^{-5}
12) 2×10^{-5}
13) 2.7×10^{-4}
14) 7×10^4
15) 2.87×10^3
16) 1.9×10^5
17) 2.23×10^{-2}
18) 7×10^{-1}
19) 8.2×10^{-2}
20) 3.1×10^5
21) 4.8×10^4
22) 9.8×10^{-5}
23) 3.5×10^{-2}
24) 1.778×10^3
25) 5.8781×10^4
26) 2.45×10^4
27) 3.3021×10^4
28) 8.1×10^6

Square Roots

1) 8
2) 0
3) 18
4) 22
5) 40
6) 23
7) 0.1
8) 100
9) 0.4
10) 0.6
11) 0.5
12) 1.1
13) 28
14) 24
15) 26
16) 31
17) 41
18) 0.9
19) 0.7
20) 0.8
21) 33
22) 50
23) 90
24) 110
25) 1.5
26) 1.3
27) 1.2
28) 0.2

Simplify Square Roots

1) $3\sqrt{6}$
2) $6\sqrt{3}$
3) $2\sqrt{3}$
4) $3\sqrt{11}$
5) $10\sqrt{2}$
6) $3\sqrt{5}$
7) $40\sqrt{2}$
8) $30\sqrt{3}$
9) $2\sqrt{6}$
10) $6\sqrt{2}$
11) $11\sqrt{3}$
12) $4 - \sqrt{5}$
13) $4\sqrt{3}$
14) $3 + \sqrt{5}$
15) 6
16) 10
17) 1
18) $4y^3\sqrt{5}$
19) $54\sqrt{a}$
20) 10
21) $6\sqrt{2}$
22) $12\sqrt{3}$
23) $4\sqrt{7}$
24) $8\sqrt{2}$
25) $16\sqrt{3}$
26) $4\sqrt{6}$

Chapter 7: Ratio, Proportion and Percent

Proportions

Find a missing number in a proportion.

1) $\dfrac{4}{7} = \dfrac{12}{a}$

2) $\dfrac{a}{9} = \dfrac{20}{45}$

3) $\dfrac{12}{60} = \dfrac{a}{5}$

4) $\dfrac{16}{a} = \dfrac{96}{36}$

5) $\dfrac{4}{a} = \dfrac{16}{75}$

6) $\dfrac{\sqrt{9}}{4} = \dfrac{a}{32}$

7) $\dfrac{2}{4} = \dfrac{18}{a}$

8) $\dfrac{7}{14} = \dfrac{a}{35}$

9) $\dfrac{7}{a} = \dfrac{4.2}{6}$

10) $\dfrac{2}{12} = \dfrac{8}{a}$

11) $\dfrac{10}{6} = \dfrac{5}{a}$

12) $\dfrac{16}{a} = \dfrac{4}{19}$

13) $\dfrac{4}{11} = \dfrac{a}{12}$

14) $\dfrac{\sqrt{36}}{3} = \dfrac{48}{a}$

15) $\dfrac{6}{a} = \dfrac{6.6}{39.6}$

16) $\dfrac{60}{140} = \dfrac{a}{280}$

17) $\dfrac{42}{200} = \dfrac{a}{68}$

18) $\dfrac{26}{104} = \dfrac{a}{4}$

19) $\dfrac{10}{16} = \dfrac{2}{a}$

20) $\dfrac{9}{7} = \dfrac{27}{a}$

Reduce Ratio

Reduce each ratio to the simplest form.

1) $5:20 =$

2) $6:36 =$

3) $63:35 =$

4) $24:20 =$

5) $12:120 =$

6) $16:2 =$

7) $70:350 =$

8) $4:144 =$

9) $25:75 =$

10) $4.8:5.6 =$

11) $110:330 =$

12) $3:5 =$

13) $120:200 =$

14) $30:45 =$

15) $34:68 =$

16) $32:8 =$

17) $140:35 =$

18) $20:200 =$

19) $126:84 =$

20) $156:198 =$

21) $40:80 =$

22) $42:49 =$

23) $5:75 =$

24) $18:108 =$

Word Problems

Solve each word problem.

1) Bob has 16 red cards and 40 green cards. What is the ratio of Bob's red cards to his green cards? _____

2) In a party, 20 soft drinks are required for every 24 guests. If there are 504 guests, how many soft drinks is required? _____

3) In Bob's class, 36 of the students are tall and 20 are short. In Mason's class 108 students are tall and 60 students are short. Which class has a higher ratio of tall to short students? _____

4) The price of 4 apples at the Quick Market is $1.92. The price of 10 of the same apples at Walmart is $5.00. Which place is the better buy? _____

5) The bakers at a Bakery can make 250 bagels in 5 hours. How many bagels can they bake in 12 hours? What is that rate per hour? _____

6) You can buy 15 cans of green beans at a supermarket for $9.20. How much does it cost to buy 45 cans of green beans? _____

7) The ratio of boys to girls in a class is 5:4. If there are 15 boys in the class, how many girls are in that class? _____

8) The ratio of red marbles to blue marbles in a bag is 2:5. If there are 49 marbles in the bag, how many of the marbles are red? _____

Percent

Find the Percent of Numbers.

1) 30% of 42 =

2) 28% of 15 =

3) 15% of 14 =

4) 24% of 70 =

5) 8% of 80 =

6) 35% of 12 =

7) 18% of 5 =

8) 12% of 46 =

9) 40% of 62 =

10) 4.5% of 50 =

11) 85% of 18 =

12) 60% of 50 =

13) 18% of 180 =

14) 2% of 240 =

15) 75% of 0 =

16) 80% of 120 =

17) 36% of 45 =

18) 10% of 70 =

19) 8% of 13 =

20) 4% of 8 =

21) 30% of 44 =

22) 80% of 17 =

23) 22% of 35 =

24) 8% of 150 =

25) 40% of 270 =

26) 2% of 5 =

27) 9% of 320 =

28) 10% of 26 =

Convert Fraction to Percent

Write each as a percent.

1) $\frac{1}{4} =$

2) $\frac{3}{8} =$

3) $\frac{7}{14} =$

4) $\frac{15}{35} =$

5) $\frac{12}{28} =$

6) $\frac{17}{68} =$

7) $\frac{8}{11} =$

8) $\frac{14}{30} =$

9) $\frac{6}{50} =$

10) $\frac{12}{48} =$

11) $\frac{5}{34} =$

12) $\frac{27}{10} =$

13) $\frac{24}{80} =$

14) $\frac{16}{25} =$

15) $\frac{16}{58} =$

16) $\frac{2}{22} =$

17) $\frac{32}{88} =$

18) $\frac{21}{36} =$

19) $\frac{18}{92} =$

20) $\frac{6}{60} =$

21) $\frac{24}{600} =$

22) $\frac{720}{360} =$

Convert Decimal to Percent

Write each as a percent.

1) 0.187 =

2) 0.19 =

3) 2.6 =

4) 0.017 =

5) 0.009 =

6) 0.786 =

7) 0.245 =

8) 0.57 =

9) 0.002 =

10) 0.205 =

11) 0.324 =

12) 84.9 =

13) 3.015 =

14) 0.7 =

15) 2.35 =

16) 0.0367 =

17) 0.0043 =

18) 0.960 =

19) 6.68 =

20) 0.484 =

21) 8.957 =

22) 0.879 =

23) 2.7 =

24) 0.9 =

25) 3.6 =

26) 26.8 =

27) 1.01 =

28) 0.006 =

Answer key Chapter 7

Proportions

1) $a = 21$
2) $a = 4$
3) $a = 1$
4) $a = 6$
5) $a = 18.75$
6) $a = 24$
7) $a = 36$
8) $a = 17.5$
9) $a = 10$
10) $a = 48$
11) $a = 3$
12) $a = 76$
13) $a = \frac{48}{11}$
14) $a = 24$
15) $a = 36$
16) $a = 120$
17) $a = 14.28$
18) $a = 1$
19) $a = 3.2$
20) $a = 21$

Reduce Ratio

1) $1:4$
2) $1:6$
3) $9:5$
4) $6:5$
5) $1:10$
6) $8:1$
7) $1:5$
8) $1:36$
9) $1:3$
10) $0.6:0.7$
11) $11:33$
12) $0.6:1$
13) $3:5$
14) $2:3$
15) $1:2$
16) $4:1$
17) $4:1$
18) $1:10$
19) $3:2$
20) $26:33$
21) $1:2$
22) $6:7$
23) $1:15$
24) $1:6$

Word Problems

1) 2:5
2) 420
3) The ratio for both classes is 9 to 5.
4) Quick Market is a better buy.
5) 600, the rate is 50 per hour.
6) $27.60
7) 12
8) 14

Percent

1) 12.6
2) 4.2
3) 2.1
4) 16.8
5) 6.4
6) 4.2
7) 0.9
8) 5.52
9) 24.8
10) 2.25
11) 15.3
12) 30
13) 13.4
14) 4.8
15) 0

16) 96
17) 16.2
18) 7
19) 1.04
20) 0.32

21) 13.2
22) 13.6
23) 7.7
24) 12
25) 108

26) 0.1
27) 28.8
28) 2.6

Convert Fraction to Percent

1) 25%
2) 37.5%
3) 50%
4) 42.86%
5) 29.31%
6) 25%
7) 72.72%
8) 46.66%

9) 12%
10) 25%
11) 14.7%
12) 2.7%
13) 30%
14) 64%
15) 27.58%
16) 9.09%

17) 36.36%
18) 58.33%
19) 19.56%
20) 10%
21) 4%
22) 200%

Convert Decimal to Percent

1) 18.7%
2) 19%
3) 260%
4) 1.7%
5) 0.9%
6) 78.6%
7) 24.5%
8) 57%
9) 0.2%
10) 20.5%

11) 32.4%
12) 8,490%
13) 301.5%
14) 70%
15) 235%
16) 3.67%
17) 0.43%
18) 96%
19) 668%
20) 48.4%

21) 895.7%
22) 87.9%
23) 270%
24) 90%
25) 360%
26) 2,680%
27) 101%
28) 0.6%

Chapter 8:

Measurement

Reference Measurement

LENGTH	
Customary	**Metric**
1 mile (mi) = 1,760 yards (yd)	1 kilometer (km) = 1,000 meters (m)
1 yard (yd) = 3 feet (ft)	1 meter (m) = 100 centimeters (cm)
1 foot (ft) = 12 inches (in.)	1 centimeter(cm) = 10 millimeters(mm)
VOLUME AND CAPACITY	
Customary	**Metric**
1 gallon (gal) = 4 quarts (qt)	1 liter (L) = 1,000 milliliters (mL)
1 quart (qt) = 2 pints (pt.)	
1 pint (pt.) = 2 cups (c)	
1 cup (c) = 8 fluid ounces (Fl oz)	
WEIGHT AND MASS	
Customary	**Metric**
1 ton (T) = 2,000 pounds (lb.)	1 kilogram (kg) = 1,000 grams (g)
1 pound (lb.) = 16 ounces (oz)	1 gram (g) = 1,000 milligrams (mg)
Time	
1 year = 12 months	
1 year = 52 weeks	
1 week = 7 days	
1 day = 24 hours	
1 hour = 60 minutes	
1 minute = 60 seconds	

Common Core Math Workbook

Metric Length Measurement

Convert to the units.

1) 5×10^4 mm = _____ cm

2) 0.4 m = _____ mm

3) 0.06 m = _____ cm

4) 1.2 km = _____ m

5) 8,000 mm = _____ m

6) 4,700 cm = _____ m

7) 4.5 m = _____ cm

8) 7×10^3 mm = _____ cm

9) 9×10^6 mm = _____ m

10) 2 km = _____ mm

11) 0.3 km = _____ m

12) 0.05 m = _____ cm

13) 4×10^4 m = _____ km

14) 6×10^7 m = _____ km

Customary Length Measurement

Convert to the units.

1) 20 ft = _____ in

2) 2.5 ft = _____ in

3) 5.6 yd = _____ ft

4) 0.4 yd = _____ ft

5) 9×10^{-1} yd = _____ in

6) 2 mi = _____ in

7) 18×10^3 in = _____ yd

8) 21.6 in = _____ yd

9) 6,160 yd = _____ mi

10) 28 yd = _____ in

11) 0.03 mi = _____ yd

12) 99×10^3 ft = _____ mi

13) 4.8 in = _____ ft

14) 42 yd = _____ feet

15) 0.72 in = _____ ft

16) 0.2 mi = _____ ft

Metric Capacity Measurement

Convert the following measurements.

1) 60 l = _____ ml

2) 0.7 l = _____ ml

3) 2.8 l = _____ ml

4) 0.06 l = _____ ml

5) 22.5 l = _____ ml

6) 0.9 l = _____ ml

7) 6×10^6 ml = _____ l

8) 22×10^5 ml = _____ l

9) 112×10^2 ml = _____ l

10) 11,000 ml = _____ l

11) 57,800 ml = _____ l

12) 0.3×10^5 ml = _____ l

Customary Capacity Measurement

Convert the following measurements.

1) 1.5 gal = _____ qt.

2) 4.5 gal = _____ pt.

3) 0.5 gal = _____ c.

4) 18 pt. = _____ c

5) 12 c = _____ fl oz

6) 8.15 qt = _____ pt.

7) 0.08 qt = _____ c

8) 42 pt. = _____ c

9) 8×10^4 c = _____ gal

10) 256 pt. = _____ gal

11) 484 qt = _____ gal

12) 25.8 pt. = _____ qt

13) 7×10^3 c = _____ qt

14) 98.8 c = _____ pt.

15) 0.164 qt = _____ gal

16) 1,256 pt. = _____ qt

17) 23 gal = _____ pt.

18) 0.01 qt = _____ c

19) 800 c = _____ gal

20) 64.16 fl oz = _____ c

Common Core Math Workbook

Metric Weight and Mass Measurement

Convert.

1) 0.8 kg = _____ g

2) 5.6 kg = _____ g

3) 2×10^{-4} kg = _____ g

4) 1.04 kg = _____ g

5) 44.8 kg = _____ g

6) 13.12 kg = _____ g

7) 0.072 kg = _____ g

8) 21×10^5 g = _____ kg

9) 15×10^6 g = _____ kg

10) 0.04×10^8 g = _____ kg

11) 17,400 g = _____ kg

12) 98×10^2 g = _____ kg

13) 5,400,000 g = _____ kg

14) 325×10^4 g = _____ kg

Customary Weight and Mass Measurement

Convert.

1) 24×10^4 lb. = _____ T

2) 0.32×10^5 lb. = _____ T

3) 190,000 lb. = _____ T

4) 2,800 lb. = _____ T

5) 0.35 lb. = _____ oz

6) 2.8 lb. = _____ oz

7) 0.05 lb. = _____ oz

8) 4 T = _____ lb.

9) 7×10^{-4} T = _____ lb.

10) 38×10^{-5} T = _____ lb.

11) 0.6 T = _____ lb.

12) 0.003 T = _____ oz

13) 0.015 T = _____ oz

14) 196.8 oz = _____ lb.

WWW.MathNotion.com

Time

Convert to the units.

1) 28 hr. = _____ min

2) 15 year = _____ week

3) 0.5 hr. = _____ sec

4) 8.5 min = _____ sec

5) 6×10^4 min = _____ hr

6) 1,095 day = _____ year

7) 2 year = _____ hr.

8) 42 day = _____ hr

9) 2 day = _____ min

10) 480 min = _____ hr

11) 28.5 year = _____ month

12) 12,600 sec = _____ min

13) 216 hr = _____ day

14) 15 weeks = _____ day

How much time has passed?

15) From 3:35 A.M. to 6:45 A.M.: ____ hours and ____ minutes.

16) From 2:30 A.M. to 7:15 A.M.: ____ hours and ____ minutes.

17) It's 6:20 P.M. What time was 3 hours ago? _____ O'clock

18) 4:15 A.M to 7:35 AM: _____ hours and _____ minutes.

19) 1:45 A.M to 4:20 AM: _____ hours and _____ minutes.

20) 9:00 A.M. to 10:05 AM. = _____ hour(s) and _____ minutes.

21) 10:35 A.M. to 3:05 PM. = _____ hour(s) and _____ minutes

22) 5:12 A.M. to 5:48 A.M. = _____ minutes

23) 8:08 A.M. to 8:45 A.M. = _____ minutes

Answers of Worksheets – Chapter 8

Metric length

1) 5,000 cm
2) 400 mm
3) 6 cm
4) 1,200 m
5) 8 m
6) 47 m
7) 450 cm
8) 700 cm
9) 9,000 m
10) 2,000,000 mm
11) 300 m
12) 5 cm
13) 40 km
14) 60,000 km

Customary Length

1) 240
2) 30
3) 16.8
4) 1.2
5) 32.4
6) 126,720
7) 500
8) 0.6
9) 3.5
10) 1,008
11) 52.8
12) 18.75
13) 0.4
14) 126
15) 0.06
16) 1,056

Metric Capacity

1) 60,000 ml
2) 700 ml
3) 2,800 ml
4) 60 ml
5) 22,500 ml
6) 900 ml
7) 6,000 ml
8) 2,200 ml
9) 11.2 ml
10) 11 L
11) 57.8 L
12) 30 L

Customary Capacity

1) 6 qt
2) 36 pt.
3) 8 c
4) 36 c
5) 96 fl oz
6) 16.3 pt.
7) 0.32 c
8) 84 c
9) 5,000 gal
10) 32 gal
11) 121 gal
12) 12.9 qt
13) 1,750 qt
14) 49.4 pt.
15) 0.041 gal
16) 628 qt
17) 184 pt.
18) 0.04 c
19) 50 gal
20) 8.02 c

Metric Weight and Mass

1) 800 g
2) 5,600 g
3) 0.2 g
4) 1,040 g
5) 44,800 g
6) 13,120 g
7) 72 g
8) 2,100 kg
9) 15,000 kg
10) 4,000 kg
11) 17.4 kg
12) 9.8 kg
13) 5,400 kg
14) 3,250 kg

Customary Weight and Mass

1) 120 T
2) 16 T
3) 95 T
4) 1.4 T
5) 5.6 oz
6) 44.8 oz
7) 0.8 oz
8) 8,000 lb.
9) 1.4 lb.
10) 0.76 lb.
11) 1,200 lb.
12) 96 oz
13) 480 oz
14) 12.3 lb

Time

1) 1,680 min
2) 780 weeks
3) 1,800 sec
4) 510 sec
5) 1,000 hr
6) 3 year
7) 17,520 hr
8) 1,008 hr
9) 2,880 min
10) 8 hr
11) 342 months
12) 210 min
13) 9 days
14) 105 days
15) 3:10
16) 4:45
17) 3:20 P.M.
18) 3:20
19) 2:35
20) 1:05
21) 4:30
22) 36 minutes
23) 37 minutes

Chapter 9:
Algebraic
Expressions

Find a Rule

Complete the output.

1- **Rule:** the output is $x + 25$

Input	x	8	15	20	38	40
Output	y					

2- **Rule:** the output is $x \times 18$

Input	x	3	7	10	11	15
Output	y					

3- **Rule:** the output is $x \div 7$

Input	x	126	147	105	280	455
Output	y					

Find a rule to write an expression.

4- **Rule:** _____

Input	x	11	13	15	20
Output	y	55	65	75	100

5- **Rule:** _____

Input	x	10	28	32	46
Output	y	14	32	36	50

6- **Rule:** _____

Input	x	84	132	180	252
Output	y	14	22	30	42

Variables and Expressions

Write a verbal expression for each algebraic expression.

1) $2a - 4b$

2) $8c^2 + 2d$

3) $x - 8$

4) $\frac{80}{15}$

5) $a^2 + b^3$

6) $2x + 4$

7) $x^2 - 10y + 8$

8) $x^3 + 9y^2 - 4$

9) $\frac{1}{3}x + \frac{3}{4}y - 6$

10) $\frac{1}{5}(x + 8) - 10y$

Write an algebraic expression for each verbal expression.

11) 9 less than h

12) The product of 9 and b

13) The 26 divided by k

14) The product of 5 and the third power of x

15) 10 more than h to the fifth power

16) 20 more than twice d

17) One fourth the square of b

18) The difference of 23 and 4 times a number

19) 60 more than the cube of a number

20) Three-quarters the cube of a number

Translate Phrases

Write an algebraic expression for each phrase.

1) A number increased by sixty–one.

2) The sum of twenty and 2 times a number

3) The difference between fifty–seven and a number.

4) The quotient of twenty-two and a number.

5) Twice a number decreased by 50.

6) four times the sum of a number and − 20.

7) A number divided by − 12.

8) The quotient of 49 and the product of a number and − 12.

9) ten subtracted from 2 times a number.

10) The difference of eight and a number.

Distributive Property

Multiply using the distributive property.

1) $5(x + 8) =$ _____

2) $2(x + 9) =$ _____

3) $(x + 4)6 =$ _____

4) $3(x + 5) =$ _____

5) $9(x + 7) =$ _____

6) $12(x + 3) =$ _____

7) $11(x + 2) =$ _____

8) $8(x + 9) =$ _____

9) $9(x + 9) =$ _____

10) $(x + 7)7 =$ _____

11) $(x + 10)5 =$ _____

12) $2(x + 13) =$ _____

13) $3(5x - 7) =$ _____

14) $4(6x - 5) =$ _____

15) $6(5x - 4) =$ _____

16) $(3x - 9)2 =$ _____

17) $(9x - 3)6 =$ _____

18) $(8x - 4)9 =$ _____

19) $5(7x - 6) =$ _____

20) $(-3)(8x - 8) =$ _____

21) $(-4)(x - 11) =$ _____

22) $(-9)(5x - 2) =$ _____

23) $(6x + 5)(-8) =$ _____

24) $(x + 8)(-11) =$ _____

Evaluate One Variable Expressions

Evaluate each using the values given.

1) $x + 4x, x = 3$

2) $5(-6 + 3x), x = 1$

3) $4x + 7x, x = -3$

4) $5(2 - x) + 5, x = 3$

5) $6x + 4x - 10, x = 2$

6) $5x + 11x + 12, x = -1$

7) $5x - 2x - 4, x = 5$

8) $\frac{3(5x+8)}{9}, x = 2$

9) $2x - 85, x = 32$

10) $\frac{x}{18}, x = 108$

11) $7(3 + 2x) - 33, x = 5$

12) $7(x + 3) - 23, x = 4$

13) $\frac{x+(-6)}{-3}, x = -6$

14) $8(6 - 3x) + 5, x = 2$

15) $-11 - \frac{x}{5} + 3x, x = 10$

16) $5x + 11x, x = 1$

17) $-12x + 3(5 + 3x), x = -7$

18) $x + 11x, x = 0.5$

19) $\frac{(2x-2)}{6}, x = 13$

20) $3(-1 - 2x), x = 5$

21) $5x - (5 - x), x = 3$

22) $\left(-\frac{15}{x}\right) + 2 + x, x = 5$

23) $-\frac{x \times 5}{x}, x = 5$

24) $2(-1 - 3x), x = 2$

25) $2x^2 + 7x, x = 1$

26) $2(3x + 1) - 4(x - 5), x = 3$

27) $-6x - 4, x = -5$

28) $7x + 2x, x = 3$

Answer key Chapter 9

Find a Rule

1)

Input	x	8	15	20	38	40
Output	y	33	40	45	63	65

2)

Input	x	3	7	10	11	15
Output	y	54	126	180	198	270

3)

Input	x	126	147	105	280	455
Output	y	18	21	15	40	65

4) y = 5x   5) y = x + 4   6) y = x ÷ 6

Variables and Expressions

1) 2 times a minus 4 times b
2) 8 times c squared plus 2 times d
3) a number minus 8
4) the quotient of 80 and 15
5) a squared plus b cubed
6) the product of 2 and x plus 4
7) x squared minus the product of 10 and y plus 8
8) x cubed plus the product of 9 and y squared minus 4
9) the sum of one–thirds of x and three–quarters of y, minus 6
10) one–fifth of the sum of x and 8 minus the product of 10 and y

11) 9 < h
12) 9b
13) $\frac{26}{K}$
14) $5x^3$
15) $10 > h^5$
16) $2d < 20$
17) $\frac{1}{4}b^2$
18) $23 - 4a$
19) $60 > a^3$
20) $\frac{3}{4}x^3$

Translate Phrases

1) x + 61
2) 20 + 2x
3) 57 − x
4) $\frac{22}{x}$
5) 2x − 50
6) 4(x + (−20))
7) $\frac{x}{-12}$
8) $\frac{49}{-12x}$
9) 2x − 10
10) 8 − x

Distributive Property

1) 5x + 40
2) 2x + 18
3) 6x + 24
4) 3x + 15
5) 9x + 63
6) 12x + 36

7) $11x + 22$
8) $8x + 72$
9) $9x + 81$
10) $7x + 49$
11) $5x + 50$
12) $2x + 26$

13) $15x - 21$
14) $24x - 20$
15) $30x - 24$
16) $6x - 18$
17) $54x - 18$
18) $72x - 36$

19) $35x - 30$
20) $-24x + 24$
21) $-4x + 44$
22) $-45x + 18$
23) $-48x - 40$
24) $-11x - 88$

Evaluate One Variable Expressions

1) 15
2) −15
3) −33
4) 0
5) 10
6) −4
7) 11

8) 6
9) −21
10) 6
11) 58
12) 26
13) 4
14) 5

15) 17
16) 16
17) 36
18) 6
19) 4
20) −33
21) 13

22) 4
23) −5
24) −14
25) 9
26) 28
27) 26
28) 27

Chapter 10: Symmetry and Transformations

Line Segments

Write each as a line, ray or line segment.

1)

2)

3)

4)

5)

6)

7)

8)

Parallel, Perpendicular and Intersecting Lines

State whether the given pair of lines are parallel, perpendicular, or intersecting.

1)

2)

3)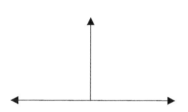

4)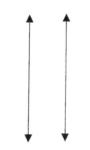

5)

6)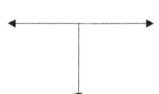

7)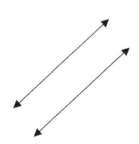

8)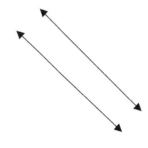

Identify Lines of Symmetry

Tell whether the line on each shape a line of symmetry is.

1)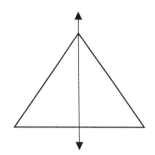

2)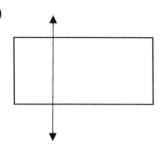

3)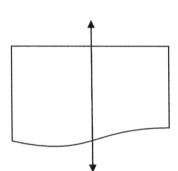

4)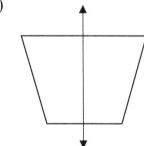

5)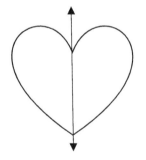

6)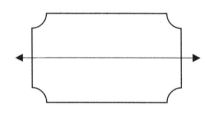

7)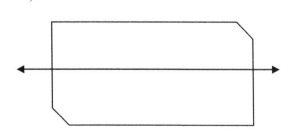

8)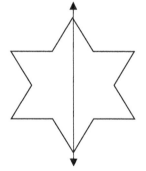

Lines of Symmetry

Draw lines of symmetry on each shape. Count and write the lines of symmetry you see.

1)

2)

3)

4)

5)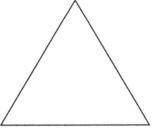

6)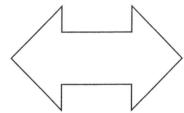

7)

8)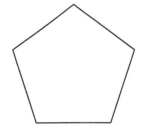

Identify Three–Dimensional Figures

Write the name of each shape.

1)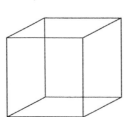

2)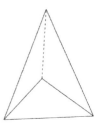

3)

4)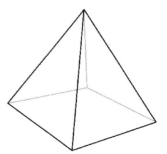

5)

6)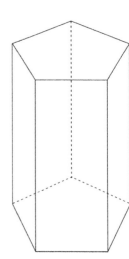

7)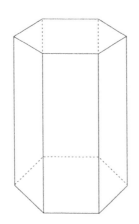

Vertices, Edges, and Faces

Complete the chart below.

	Shape	Number of edges	Number of faces	Number of vertices
1)	triangular pyramid	_____	_____	_____
2)	square pyramid	_____	_____	_____
3)	cube	_____	_____	_____
4)	rectangular prism	_____	_____	_____
5)	pentagonal prism	_____	_____	_____
6)	hexagonal prism	_____	_____	_____

Identify Faces of Three–Dimensional Figures

Write the number of faces.

1)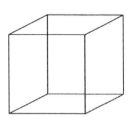

2)

3)

4)

5)

6)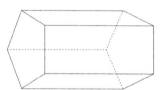

7)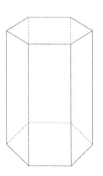

8)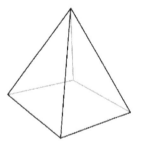

Answers of Worksheets – Chapter 10

Line Segments

1) Line segment
2) Ray
3) Line
4) Line segment
5) Ray
6) Line
7) Line
8) Line segment

Parallel, Perpendicular and Intersecting Lines

1) Parallel
2) Intersection
3) Perpendicular
4) Parallel
5) Intersection
6) Perpendicular
7) Parallel
8) Parallel

Identify lines of symmetry

1) yes
2) no
3) no
4) yes
5) yes
6) yes
7) no
8) yes

lines of symmetry

1)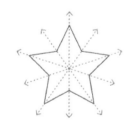

2)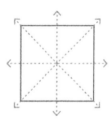

3)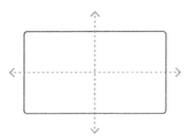

4)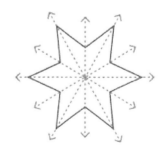

5)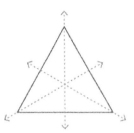
6)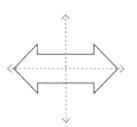
7)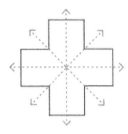
8)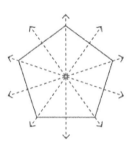

Identify Three–Dimensional Figures

1) Cube
2) Triangular pyramid
3) Triangular prism
4) Square pyramid
5) Rectangular prism
6) Pentagonal prism
7) Hexagonal prism

Vertices, Edges, and Faces

	Shape	Number of edges	Number of faces	Number of vertices
1)		6	4	4
8)		8	5	5
9)		12	6	8

Common Core Math Workbook

10)    12   6   8

11)    15   7   10

12)    18   8   12

Identify Faces of Three–Dimensional Figures

1) 6   4) 4   7) 8
2) 2   5) 6   8) 5
3) 5   6) 7

Chapter 11:

Geometry

Area and Perimeter of Square

Find the perimeter and area of each squares.

1)

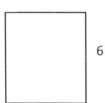

Perimeter: _____
Area: _____

2)

Perimeter: _____
Area: _____

3)

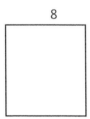

Perimeter: _____
Area: _____

4)

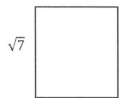

Perimeter: _____
Area: _____

5)

Perimeter: _____
Area: _____

6)

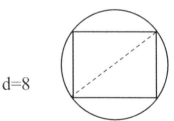

Perimeter of Square: _____
Area of Square: _____

Area and Perimeter of Rectangle

Find the perimeter and area of each rectangle.

1)

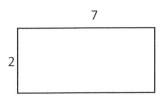

Perimeter: _____
Area: _____

2)

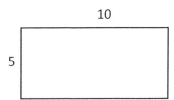

Perimeter: _____
Area: _____

3)

13
7

Perimeter: _____
Area: _____

4)

8
1.5

Perimeter: _____
Area: _____

5)

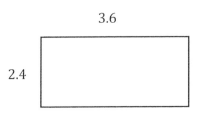

Perimeter: _____
Area: _____

6)

6
4

Perimeter: _____
Area: _____

Area and Perimeter of Triangle

Find the perimeter and area of each triangle.

1)

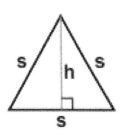

Perimeter: _____.

Area: _____.

2)

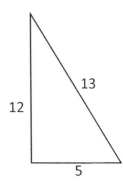

Perimeter: _____.

Area: _____.

3)

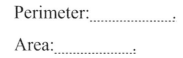

Perimeter: _____.

Area _____.

4)

s=12

h=8

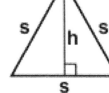

Perimeter: _____.

Area: _____.

5)

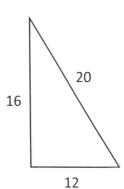

Perimeter: _____.

Area: _____.

6)

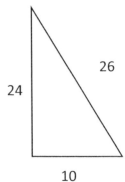

Perimeter: _____.

Area: _____.

Area and Perimeter of Trapezoid

Find the perimeter and area of each trapezoid.

1)
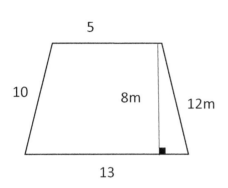

2)
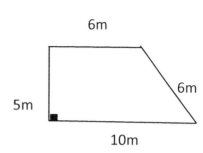

Perimeter:_____:

Area:_____:

Perimeter:_____:

Area:_____:

3)
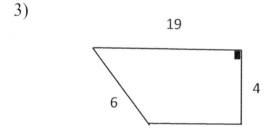

4)
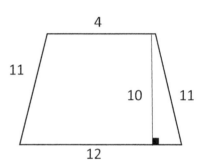

Perimeter:_____.

Area _____:

Perimeter:_____:

Area:_____:

5)
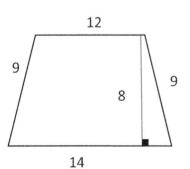

6)
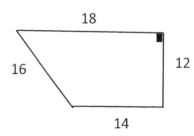

Perimeter:_____.

Area:_____:

Perimeter:_____:

Area:_____:

Area and Perimeter of Parallelogram

Find the perimeter and area of each parallelogram.

1)

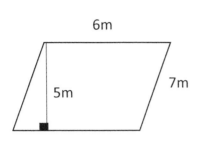

Perimeter: _____.

Area: _____.

2)

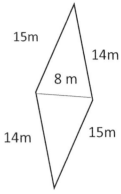

Perimeter: _____.

Area: _____.

3)

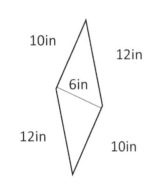

Perimeter: _____.

Area _____.

4)

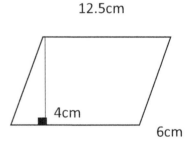

Perimeter: _____.

Area: _____.

5)

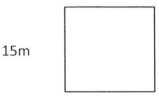

Perimeter: _____.

Area: _____.

6)

Perimeter: _____.

Area: _____.

Circumference and Area of Circle

Find the circumference and area of each ($\pi = 3.14$).

1)

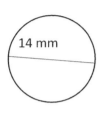

Circumference:

Area:

2)

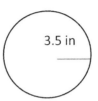

Circumference:

Area:

3)

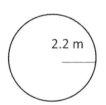

Circumference:

Area

4)

Circumference:

Area:

5)

Circumference:

Area:

6)

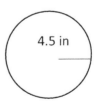

Circumference:

Area:

Perimeter of Polygon

Find the perimeter of each polygon.

1)
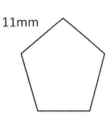
11mm

Perimeter:_____.

2)
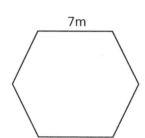
7m

Perimeter:_____:

3)
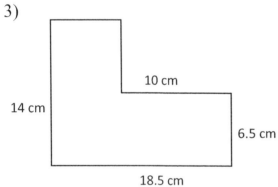
14 cm, 10 cm, 6.5 cm, 18.5 cm

Perimeter:_____.

4)
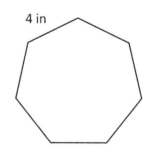
4 in

Perimeter:_____:

5)
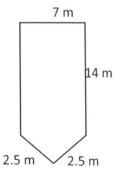
7 m, 14 m, 2.5 m, 2.5 m

Perimeter:_____.

6)
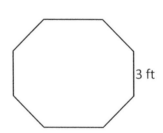
3 ft

Perimeter:_____:

Volume of Cubes

Find the volume of each cube.

1)
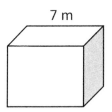

V: _____.

2)
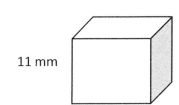

V: _____.

3)
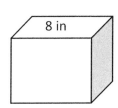

V: _____.

4)
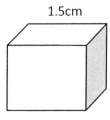

V: _____.

5)
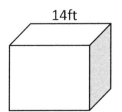

V: _____.

6)
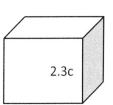

V: _____.

Volume of Rectangle Prism

Find the volume of each rectangle prism

1)

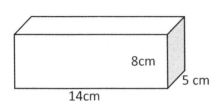

V:_____.

2)

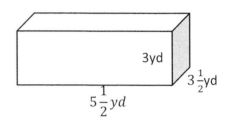

V:_____.

3)

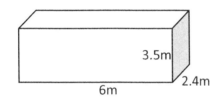

V:_____.

4)

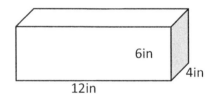

V:_____.

5)

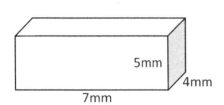

V:_____.

6)

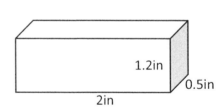

V:_____.

Answer key Chapter 11

Area and Perimeter of Square

1. Perimeter: 24,   Area: 36
2. Perimeter: $4\sqrt{5}$, Area: 5
3. Perimeter: 32,   Area: 64
4. Perimeter: $4\sqrt{7}$, Area: 7
5. Perimeter: 44,   Area: 121
6. Perimeter: $4\sqrt{32}$, Area: 32

Area and Perimeter of Rectangle

1- Perimeter: 18,   Area: 14
2- Perimeter: 30,   Area: 50
3- Perimeter: 40,   Area: 91
4- Perimeter: 19, Area: 12
5- Perimeter: 12, Area: 8.64
6- Perimeter: 20,   Area: 24

Area and Perimeter of Triangle

1- Perimeter: 3s, Area: $\frac{1}{2}sh$
2- Perimeter: 30,   Area: 30
3- Perimeter: 32,   Area: 32
4- Perimeter: 36,   Area: 48
5- Perimeter: 48,   Area: 96
6- Perimeter: 60, Area: 120

Area and Perimeter of Trapezoid

1- Perimeter: 40,   Area: 72
2- Perimeter: 27,   Area: 40
3- Perimeter: 41,   Area: 62
4- Perimeter: 38,   Area: 80
5- Perimeter: 44,   Area: 104
6- Perimeter: 60,   Area: 192

Area and Perimeter of Parallelogram

1- Perimeter: $26m$, Area: $30(m)^2$
2- Perimeter: $58m$, Area: $120(m)^2$
3- Perimeter: $44in$, Area: $60(in)^2$
4- Perimeter: $37cm$, Area: $50(cm)^2$
5- Perimeter: $80m$, Area: $364(m)^2$
6- Perimeter: $60m$, Area: $225(m)^2$

Circumference and Area of Circle

1) Circumference: 43.96 mm   Area: $153.86(mm)^2$
2) Circumference: 21.98 in   Area: $38.465(in)^2$
3) Circumference: 13.816 m   Area: $15.197(m)^2$
4) Circumference: 31.4 cm   Area: $78.5(cm)^2$
5) Circumference: 18.84 in   Area: $28.26(in)^2$
6) Circumference: 28.26 km   Area: $63.59(km)^2$

Perimeter of Polygon

1) 55 mm
2) 42 m
3) 65 cm
4) 28 in
5) 40 m
6) 24 ft

Volume of Cubes

1) $343 m^3$
2) $1,331 (mm)^3$
3) $512 in^3$
4) $3.375 (cm)^3$

5) $2,744 (ft)^3$ 6) $12.167 (cm)^3$

Volume of Rectangle Prism

1) $560 (cm)^3$ 3) $50.4 (m)^3$ 5) $140 (mm)^3$

2) $57.75 (yd)^3$ 4) $288 (in)^3$ 6) $1.2 (in)^3$

Chapter 12: Data and Graphs

Common Core Math Workbook

Mean and Median

Find the mean and median of the following data.

1) 24, 59, 20, 37, 14, 24, 47

Mean: __, Median: __

2) 6, 13, 13, 19, 15, 10

Mean: __, Median: __

3) 21, 35, 49, 11, 45, 27, 35, 19, 14

Mean: __, Median: __

4) 25, 11, 1, 15, 25, 18

Mean: __, Median: __

5) 24, 14, 14, 17, 23, 15, 14, 29, 29, 8

Mean: __, Median: __

6) 7, 14, 19, 11, 8, 19, 8, 15

Mean: __, Median: __

7) 29, 28, 66, 76, 14, 44, 18, 44, 22, 44

Mean: __, Median: __

8) 35, 35, 57, 78, 59

Mean: __, Median: __

9) 16, 16, 29, 46, 54

Mean: __, Median: __

10) 13, 9, 3, 3, 5, 6, 7

Mean: __, Median: __

11) 4, 12, 4, 6, 1, 8

Mean: __, Median: __

12) 8, 9, 15, 15, 17, 17, 17

Mean: __, Median: __

13) 7, 7, 1, 16, 1, 7, 19

Mean: __, Median: __

14) 13, 17, 10, 12, 12, 18, 15, 19

Mean: __, Median: __

15) 9, 14, 19, 19, 29

Mean: __, Median: __

16) 6, 6, 16, 18, 15, 22, 37

Mean: __, Median: __

17) 25, 11, 14, 25, 18, 13, 7, 5

Mean: __, Median: __

18) 55, 34, 34, 48, 85, 7

Mean: __, Median: __

19) 54, 28, 28, 65, 5, 8

Mean: __, Median: __

20) 88, 84, 23, 26, 11, 88, 19

Mean: __, Median: __

www.MathNotion.com

Mode and Range

Find the mode(s), and range of the following data.

1) 24, 59, 20, 37, 14, 24, 47

   Mode: __, Range: __

2) 6, 13, 13, 19, 15, 10

   Mode: __, Range: __

3) 21, 35, 49, 11, 45, 27, 35, 19, 14

   Mode: __, Range: __

4) 25, 11, 1, 15, 25, 18

   Mode: __, Range: __

5) 24, 14, 14, 17, 23, 15, 14, 29, 29, 8

   Mode: __, Range: __

6) 7, 14, 19, 11, 8, 19, 8, 15

   Mode: __, Range: __

7) 29, 28, 66, 76, 14, 44, 18, 44, 22, 44

   Mode: __, Range: __

8) 35, 35, 57, 78, 59

   Mode: __, Range: __

9) 16, 16, 29, 46, 54

   Mode: __, Range: __

10) 13, 9, 3, 3, 5, 6, 7

    Mode: __, Range: __

11) 4, 12, 4, 6, 1, 8

    Mode: __, Range: __

12) 8, 9, 15, 15, 17, 17, 17

    Mode: __, Range: __

13) 7, 7, 1, 16, 1, 7, 19

    Mode: __, Range: __

14) 13, 17, 10, 12, 12, 18, 15, 19

    Mode: __, Range: __

15) 9, 14, 19, 19, 29

    Mode: __, Range: __

16) 6, 6, 16, 18, 15, 22, 37

    Mode: __, Range: __

17) 25, 11, 14, 25, 18, 13, 7, 5

    Mode: __, Range: __

18) 55, 34, 34, 48, 85, 7

    Mode: __, Range: __

19) 54, 28, 28, 65, 5, 8

    Mode: __, Range: __

20) 88, 84, 23, 26, 11, 88, 19

    Mode: __, Range: __

Stem–And–Leaf Plot

Make stem-and-leaf plots for the given data.

1) 15, 16, 38, 31, 12, 54, 18, 37, 39, 34, 19, 32, 55

2) 72, 74, 17, 41, 72, 14, 46, 78, 48, 44, 49, 42

3) 125, 108, 65, 65, 105, 127, 62, 126, 68, 124, 66, 109

4) 61, 45, 66, 60, 99, 63, 90, 97, 68, 63, 49, 42

5) 55, 58, 105, 56, 15, 108, 102

6) 123, 57, 77, 55, 120, 127, 73, 124, 58, 123, 79, 71

Dot plots

The ages of students in a Math class are given below.

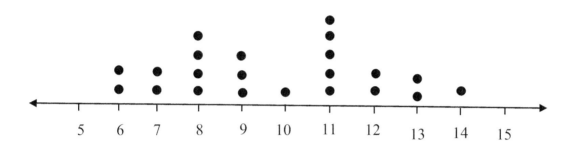

1) What is the total number of students in math class?

2) How many students are at least 12 years old?

3) Which age(s) has the most students?

4) Which age(s) has the fewest student?

5) Determine the median of the data.

6) Determine the range of the data.

7) Determine the mode of the data.

Bar Graph

Each student in class selected two games that they would like to play. Graph the given information as a bar graph and answer the questions below:

Game	Votes
Football	13
Volleyball	10
Basketball	18
Baseball	17
Tennis	13

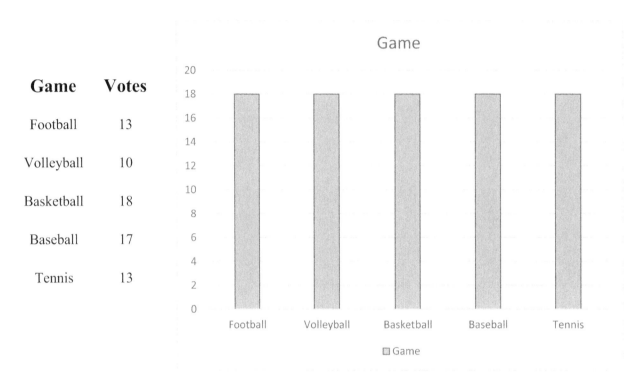

1) Which was the most popular game to play?

2) How many more students like Basketball than Volleyball?

3) Which two game got the same number of votes?

4) How many Volleyball and Football did student vote in all?

5) Did more student like football or Volleyball?

6) Which game did the fewest student like?

Probability

1) A jar contains 16 caramels, 5 mints and 19 dark chocolates. What is the probability of selecting a mint?

2) If you were to roll the dice one time what is the probability it will NOT land on a 4?

3) A die has sides are numbered 1 to 6. If the cube is thrown once, what is the probability of rolling a 5?

4) The sides of number cube have the numbers 4, 6, 8, 4, 6, and 8. If the cube is thrown once, what is the probability of rolling a 6?

5) Your friend asks you to think of a number from ten to twenty. What is the probability that his number will be 15?

6) A person has 8 coins in their pocket. 2 dime, 3 pennies, 2 quarter, and a nickel. If a person randomly picks one coin out of their pocket. What would the probability be that they get a penny?

7) What is the probability of drawing an odd numbered card from a standard deck of shuffled cards (Ace is one)?

8) 32 students apply to go on a school trip. Three students are selected at random. what is the probability of selecting 4 students?

Answer key Chapter 12

Mean, Median, Mode, and Range of the Given Data

1) mean: 32.14, median: 24
2) mean: 12.67, median: 13
3) mean: 28.44, median: 27
4) mean: 15.83, median: 16.5
5) mean: 18.7, median: 16
6) mean: 12.63, median: 12.5
7) mean: 38.5, median: 36.5
8) mean: 52.8, median: 57
9) mean: 32.2, median: 29
10) mean: 6.57, median: 6
11) mean: 5.83, median: 5
12) mean: 14, median: 15
13) mean: 8.29, median: 7
14) mean: 14.5, median: 14
15) mean: 18, median: 19
16) mean: 17.14, median: 16
17) mean: 14.75, median: 13.5
18) mean: 43.83, median: 41
19) mean: 31.33, median: 28
20) mean: 48.43, median: 26

Mean, Median, Mode, and Range of the Given Data

1) mode: 24, range: 45
2) mode: 13, range: 13
3) mode: 35, range: 38
4) mode: 25, range: 24
5) mode: 14, range: 21
6) mode: 19, 8, range: 12
7) mode: 44, range: 62
8) mode: 35, range: 43
9) mode: 16, range: 38
10) mode: 3, range: 10
11) mode: 4, range: 11
12) mode: 17, range: 9
13) mode: 7, range: 18
14) mode: 12, range: 9
15) mode: 19, range: 20
16) mode: 6, range: 31
17) mode: 25, range: 20
18) mode: 34, range: 78
19) mode: 28, range: 60
20) mode: 88, range: 77

Stem–And–Leaf Plot

1)

Stem	leaf
1	2 5 6 8 9
3	1 2 4 7 8 9
5	4 5

2)

Stem	leaf
1	4 7
4	1 2 4 6 8 9
7	2 2 4 8

3)

Stem	leaf
6	2 5 5 6 8
10	5 8 9
12	4 5 6 7

Common Core Math Workbook

4)

Stem	leaf
4	2 9 5
6	0 1 3 3 6 8
9	0 7 9

5)

Stem	leaf
1	5
5	5 6 8
10	2 5 8

6)

Stem	leaf
5	5 7 8
7	1 3 7 9
12	0 3 3 4 7

Dot plots

1) 22
2) 5
3) 11
4) 10 and 14
5) 2
6) 4
7) 2

Bar Graph

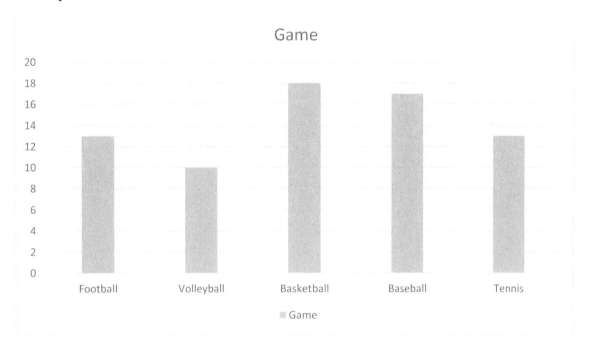

1) Basketball
2) 8 students
3) Football and Tennis
4) 23
5) Football
6) Volleyball

Probability

1) $\frac{1}{8}$
2) $\frac{5}{6}$
3) $\frac{1}{6}$
4) $\frac{1}{3}$
5) $\frac{1}{10}$
6) $\frac{3}{8}$
7) $\frac{5}{13}$
8) $\frac{1}{8}$

Common Core Test Review

Common Core GRADE 5 MAHEMATICS REFRENCE MATERIALS

Perimeter

Square $\qquad P = 4S$

Rectangle $\qquad P = 2L + 2W$

Area

Square $\qquad A = S \times S$

Rectangle $\qquad A = l \times w \qquad$ or $\qquad A = bh$

Volume

Cube $\qquad A = S \times S \times S$

Rectangular Prism $\qquad A = l \times w \times h \qquad$ or $\qquad A = Bh$

LENGTH

Customary	Metric
1 mile (mi) = 1,760 yards (yd)	1 kilometer (km) = 1,000 meters (m)
1 yard (yd) = 3 feet (ft)	1 meter (m) = 100 centimeters (cm)
1 foot (ft) = 12 inches (in.)	1 centimeter (cm) = 10 millimeters (mm)

VOLUME AND CAPACITY

Customary	Metric
1 gallon (gal) = 4 quarts (qt)	1 liter (L) = 1,000 milliliters (mL)
1 quart (qt) = 2 pints (pt.)	
1 pint (pt.) = 2 cups (c)	
1 cup (c) = 8 fluid ounces (Fl oz)	

WEIGHT AND MASS

Customary	Metric
1 ton (T) = 2,000 pounds (lb.)	1 kilogram (kg) = 1,000 grams (g)
1 pound (lb.) = 16 ounces (oz)	1 gram (g) = 1,000 milligrams (mg)

Common Core Practice Test 1

Mathematics

GRADE 5

Administered *Month Year*

1) Which expression correctly shows the sum of product of 7 and 12 and the difference of 19 and 8?

   A. $12 + (7 \times 19) - 8$

   B. $(12 \times 7) + (19 - 8)$

   C. $(12 \times 7) - (19 - 8)$

   D. $12 - (7 \times 19) + 8$

2) Which equation has the same unknown value as $528 \div 24 = \square$?

   A. $528 \times \square = 24$

   B. $\square \div 528 = 24$

   C. $24 \times \square = 528$

   D. $\square \div 24 = 528$

3) The owner of a snow-cone stands used $\frac{1}{6}$ gallon of syrup to make 18 cherry snow cones. She used the same amount of syrup in each snow cone. How much syrup in gallons was used in each cherry snow cone?

   A. $\frac{1}{3}$ gal

   B. 3 gal

   C. $\frac{1}{108}$ gal

   D. 108 gal

4) 34 students equally share a bag of 646 dimes and 680 nickels. How many dimes does each student get?

   A. 19

   B. 20

   C. 39

   D. 190

5) Bertha bought 8 cans of tuna at $1.45 a can. How much did she spend?

   A. $8.60

   B. $11.60

   C. $8.45

   D. $12.45

6) Which equation shows how to multiply $12 \times 9 \times 5$ using the associative property?

   A. $12 \times 9 \times 5 = 5 \times 9 \times 12$

   B. $(12 \times 9) + (12 \times 5) = (12 \times 5) + (12 \times 9)$

   C. $(12 \times 9) + 5 = 12 \times (9 + 5)$

   D. $(12 \times 9) \times 5 = 12 \times (9 \times 5)$

7) What is 0.84 rounded to the tenths place?

   A. 0.5

   B. 0.8

   C. 0.9

   D. 1.0

8) Each time Elijah goes to the movies he spends $15.00. Which expression shows how much he spends after going to the movies t times?

   A. $15.00 + t

   B. $15.00 – t

   C. $15.00 ÷ t

   D. $15.00 × t

9) Benjamin is using a calculator to multiply 4,716 and 40. He enters 4,716 × 400 by mistake. What can Benjamin do to correct his mistake?

   A. Subtract 360 from the product

   B. Add 360 to product

   C. Divide the product by 10

   D. Multiply the product by 10

10) The price of shoes in a store is $35 and the price of belt in the same store is $10. A customer buys 2 shoes and 3 belts during a sale when the price of shoes is discounted 25% and the price of belt is discounted 5%. How much did the customer save due to the sale?

   A. $19

   B. $9.25

   C. $32.5

   D. $31.8

11) Daniel wrote the expression shown. $24 \div 6 + 12(63 - 9)$. What do these parentheses indicate in the expression?

A. Divide 24 by 6 before adding 12

B. Multiply 12 by 63 before subtracting 9

C. Subtract 9 from 63 before multiplying by 12

D. Add 6 and 12 together before subtracting 9 from 63

12) Lucas is doing his homework. It takes him 9 minutes to do 2 problems of math and 20 minutes to read 8 pages of biology. Lucas reads the same amount of biology pages each minute. Which of these is closest to the number of minutes it takes Lucas to read each page of his biology homework?

A. 2 minutes

B. 2 ½ minutes

C. 3 minutes

D. 3 ½ minutes

13) What is the sum of angles α + β in the right triangle below?

A. 120 degrees

B. 60 degrees

C. 90 degrees

D. 180 degrees

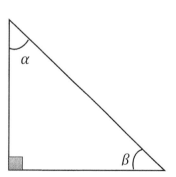

14) Which statement describe $\frac{7}{8} \times \frac{4}{9}$?

A. $\frac{7}{8} \times \frac{4}{9}$ is 7 group of $\frac{4}{9}$, divided into 72 equal parts.

B. $\frac{7}{8} \times \frac{4}{9}$ is 7 group of $\frac{4}{9}$, divided into 8 equal parts.

C. $\frac{7}{8} \times \frac{4}{9}$ is 8 group of $\frac{4}{9}$, divided into 7 equal parts.

D. $\frac{7}{8} \times \frac{4}{9}$ is 8 group of $\frac{4}{9}$, divided into 28 equal parts.

15) Noah and Liam played a new game for 7.5 hours last week. If they played the same amount of time each of 6 days, how long did they play each day?

A. 0.25 hour

B. 0.5 hour

C. 0.75 hour

D. 1.25 hour

16) Daniel planted five cucumber seeds. Out of the five planted, only three sprouted. How many plants can Daniel plan on yielding if he plants 200 seeds?

Number of seeds	5				200
Successes	3				?

A. 30

B. 60

C. 90

D. 120

Common Core Math Workbook

17) A wire is 20 feet long. The Robert needs to cut pieces that are $\frac{4}{5}$ foot long. How many pieces can he cut?

   A. 25

   B. 16

   C. 4

   D. 5

18) The temperature was 57°F at 8:00 A.M., 61°F at 8:30 A.M., and 65°F at 9:00 A.M. Describe the pattern, and predict the temperature at 11:00 A.M.

   A. Add 8°F; 97°F

   B. Add 4°F; 97°F

   C. Add 8°F; 73°F

   D. Add 4°F; 81°F

19) A rectangular prism measures 7-unit cubes wide and 4-unit cubes high. If the volume of the prism is 168 cubic unit, what is the length of the prism in unit cubes?

   A. 6

   B. 11

   C. 24

   D. 28

20) Evelyn and Mason are both saving to buy cars. So far, Mason has saved $1,062. Evelyn has saved 7 times as much as Mason. How much has Evelyn saved?

A. $7,024

B. $7,424

C. $7,434

D. $8,034

21) Which equation is true when m = 6?

A. $6 + 12 \div m = 3$

B. $12 \div m + 6 = 1$

C. $18 + 6 \div m = 4$

D. $12 \div m + 2 = 4$

22) Which is equivalent to 11 meters?

A. 0.11 cm

B. 1.1 cm

C. 110 cm

D. 1100 cm

23) If k is an odd integer, which of the following must be an even integer?

A. $k - 1$

B. $2k - 1$

C. $2k + 1$

D. $k - 2$

24) Charlotte has a grosgrain ribbon 5 feet long. She cuts the ribbon into 8 equal pieces. Which equation shows how to find the length, in feet of each piece of the ribbon?

A. $8 \times 5 = 40$

B. $5 \div 8 = \frac{5}{8}$

C. $8 + 5 = 13$

D. $5 \div 8 = 1\frac{3}{5}$

25) Paul and his four friends had lunch together. The total bill for lunch came to $33.40, including tip. If they shared the bill equally, how much did they each pay?

A. 7.85

B. 6.08

C. $6.28

D. 10.47

26) What is the volume of the figure in cubic units?

A. 28

B. 88

C. 48

D. 72

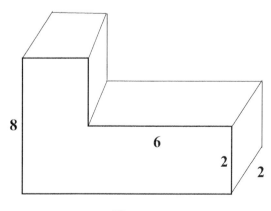

27) The graph shows three of four vertices of parallelogram ABCD. At which location on the coordinate grid could point D be located?

A. (6.5, 2)

B. (1.5, 0.5)

C. (2, 6.5)

D. (0.5, 1.5)

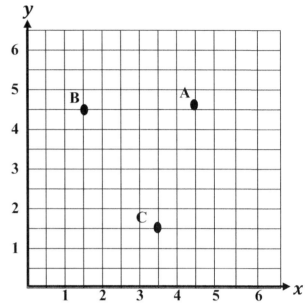

28) Which is a composite number?

A. 37

B. 47

C. 57

D. 67

29) Which capacities are written in order from greatest to least?

A. 5,215 mL, 5.21 L, 521 mL, 6 L, 6,001 mL

B. 6001 mL, 6 L, 5215 mL, 5.21 L, 521 mL

C. 521 mL, 5.21 L, 5215 mL, 6 L, 6001 mL

D. 6L, 6001 mL, 5.21 L, 5215 mL, 521 mL

30) Mrs. Smith's family ordered 12 pizzas. They ate $6\frac{2}{3}$ pizzas and gave $4\frac{5}{6}$ pizzas to her friend. How many pizzas do they have left?

A. $\frac{1}{2}$

B. $5\frac{1}{3}$

C. $7\frac{1}{6}$

D. $11\frac{1}{2}$

31) There is a big carnival every year in our town. It costs $5.00 to get in and $1.50 for every ride ticket you buy, but some of them take three-ticket rides. How much does it cost to get in and buy 12 three-ticket rides?

A. $18.50

B. $23

C. $54

D. $59

32) A cake pan, and its side lengths are shown. What does the perimeter of the cake pan in inches?

A. 12.5 in.

B. 15.2 in.

C. 25 in.

D. 18.6 in.

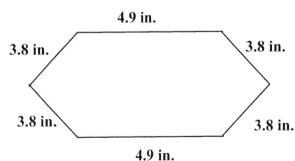

33) Eugen has a piggy bank filled with coins.

- 14 are quarters
- 19 are dimes
- 25 are nickels
- 42 are pennies

What percent of the coin are nickels?

A. 25%

B. 40%

C. 75%

D. 100%

34) Sophia is buying crullers for a group of 25 students. If 80% of the group want crullers, and each student will eat 2 crullers, how many crullers should Sophia buy?

A. 10 crullers

B. 20 crullers

C. 40 crullers

D. 80 crullers

35) Which fraction is equivalent to $\frac{3}{7}$?

A. $\frac{15}{70}$

B. $\frac{7}{3}$

C. $\frac{7}{12}$

D. $\frac{12}{28}$

36) Jayden is having a party. He wants to get two cookies for each of the 6 people, including himself, who will be at the party. If each cookie costs 60 ₵, how much money will he spend on cookies? which diagram below best shows this problem?

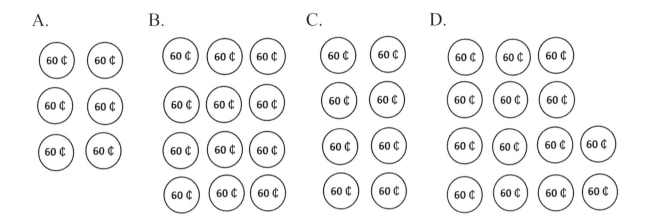

37) What is the perimeter of the rectangular in the figure below?

A. 6.2

B. 6.6

C. 8

D. 8.2

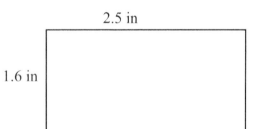

38) What is the value of C in the table below?

A. 63

B. 128

C. 149

D. 135

Input	Output
1	2
5	14
13	38
29	86
50	C

39) Robert went fishing and caught 3 fish. They weighed 19.63 ounces, 42.12 ounces, and 128.66 ounces. How much did all 3 Robert's fishes weigh together?

A. 147.36 ounces

B. 189.41 ounces

C. 190.41 ounces

D. 200.41 ounces

40) The bar graph shows the monthly high sale for a grocery store in 2018. According to the graph, for all months shown, how much smaller is the mean than the median sale?

A. 0 Million Dollars

B. 5 Million Dollars

C. 15 Million Dollars

D. 30 Million Dollars

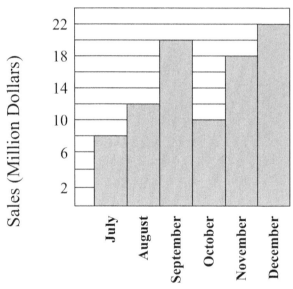

Month of Year

Common Core Practice Test 2

Mathematics

GRADE 5

Administered *Month Year*

Common Core Math Workbook

1) Which digit is the thousands digit in the number 23,090,562?

   A. 6

   B. 9

   C. 0

   D. 5

2) What is the value of expression shown? $4[4.5 - 2(1.2)]$

   A. 8.4

   B. 15.6

   C. 12

   D. 19.2

3) Which expression equal to $\frac{72}{108}$?

   A. $\frac{4}{12}$

   B. $\frac{8}{6}$

   C. $\frac{2}{8}$

   D. $\frac{2}{3}$

4) How can the distributive property be used to solve this expression? 53×24

   A. $(50 + 20) \times (3 \div 4)$

   B. $(5 \times 2) + (3 \times 4)$

   C. $(53 + 4) \times (53 + 2)$

   D. $(53 \times 20) + (53 \times 4)$

Common Core Math Workbook

5) Which statement correctly compares the two values?

   A. The value of 6 in 8.64 is 10 times the value of the 6 in 6.48

   B. The value of 6 in 8.64 is $\frac{1}{10}$ the value of the 6 in 6.48

   C. The value of 6 in 8.64 is 100 times the value of the 6 in 6.48

   D. The value of 6 in 8.64 is $\frac{1}{100}$ the value of the 6 in 6.48

6) What is 92.53 ÷ 10?

   A. 0.9253

   B. 9.253

   C. 92.53

   D. 925.3

7) Which number represents four million thirty thousand two hundred two?

   A. 4,030,022

   B. 4,300,202

   C. 4,030,202

   D. 4,300,022

8) Betty has 0.5 liters of juice. How many milliliters(mL) of juice does Betty have?

   A. 0.005 mL

   B. 0.0005 mL

   C. 5,000 mL

   D. 500 mL

WWW.MathNotion.com

9) What is sum of 7.65 and 2.8?

   A. $9\frac{29}{200}$

   B. $9\frac{73}{100}$

   C. $10\frac{1}{45}$

   D. $10\frac{9}{20}$

10) Joshua road her bicycle 26.4 miles in 4 hours. Which is how far he road in 5 minutes?

    A. 0.66 mile

    B. 0.55 miles

    C. 5.28 mile

    D. 33 miles

11) Mr. Johnson paid $36.95 for each adult shirt and $23.95 for each youth shirt he bought. Mr. Johnson bought 2 adult shirts and 5 youth shirts. How much money did he spend on these shirts?

    A. $167.35

    B. $258.65

    C. $193.65

    D. $232.65

12) Mr. Martinez had 18 daylily plants. Each plant produced 864 flowers. How many flowers did the plants produce?

A. 49

B. 882

C. 7,776

D. 15,750

13) The table below shows the lengths of different Stamps on display at a post office. Which stamps has the shortest length?

A. Stamp 1

B. Stamp 2

C. Stamp 3

D. Stamp 4

Stamp	Length (cm)
1	$\frac{1}{2}$
2	$\frac{3}{4}$
3	$\frac{3}{10}$
4	$\frac{1}{3}$

14) Ms. Bianchi works 22.5 hours a week at the movie theater. She earns $12 an hour. Which statement about her weekly income is true?

A. Her gross income is less than $270.

B. Her net income is less than $270.

C. Her gross income is more than $270.

D. Her net income is more than $270.

Common Core Math Workbook

15) A market charges $1.40 for a dozen of eggs. How many dozen eggs can you buy with $9.80?

   A. 6

   B. 7

   C. 8

   D. 9

16) An art teacher has $800. She spent $194.38 on art supplies and $478.56 on new desks. How much money does she have left?

   A. $127.06

   B. $227.60

   C. 138.06

   D. $672.94

17) Find the value of the nth term in the sequence.

Position	12	14	16	18	n
Value of Terms	4	5	6	7	

   A. $\frac{n}{3}$

   B. $n-9$

   C. $\frac{n}{2} - 2$

   D. $\frac{n}{2} + 2$

18) The dot plot shows the number of students who made from 1 to 8 words. What fraction of the student in the class made 5 or more words?

A. $\frac{1}{2}$

B. $\frac{2}{5}$

C. $\frac{3}{5}$

D. $\frac{4}{5}$

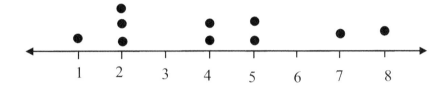

19) Jessica flips a coin 70 times. The coin lands on head 28 times. What percent of the flips were tails?

A. 28%

B. 70%

C. 40%

D. 60%

20) A parallelogram has a base of 32 cm and the height of 25 cm. a triangle has a base of 66 cm and a height of 25 cm. which figure has the greater area? By how much? (Area of parallelogram = base × high)

A. Parallelogram, 25 cm^2

B. Parallelogram, 5 cm^2

C. Triangle, 25 cm^2

D. Triangle, 35 cm^2

21) Which of the following fractions is higher than $\frac{1}{5}$, but less than $\frac{1}{2}$?

A. $\frac{3}{5}$

B. $\frac{2}{3}$

C. $\frac{1}{6}$

D. $\frac{1}{3}$

22) $\frac{13}{25}$ may be written as a percent as:

A. 13%

B. 26%

C. 44%

D. 52%

23) A chart below is showing the socks that Jenna has in her dresser. What is the probability that Jenna will choose a red sock at random?

A. $\frac{1}{3}$

B. $\frac{1}{4}$

C. $\frac{1}{5}$

D. $\frac{1}{15}$

Color	Number
Purple	3
White	2
Red	5
Stripped	4
Black	6

24) The measures of three of the interior angles of a quadrilateral are 80°, 110°, and 90°. What is the measure of the fourth angle of this quadrilateral?

   A. 30°

   B. 80°

   C. 90°

   D. 180°

25) What is the greatest common factor (GCF) of 74 and 42?

   A. 2

   B. 3

   C. 6

   D. 12

26) Which inequality is true?

   A. $0.78 > 0.87$

   B. $0.65 < 0.56$

   C. $0.45 > 0.54$

   D. $0.34 < 0.43$

27) Which percent is equivalent to 0.4?

   A. 4%

   B. 0.4%

   C. 0.004%

   D. 40%

28) James paid $12.60 each for 3 used video games and $28.5 each for 2 new video games. How much money did James spend on video games?

A. $94.8

B. $18.45

C. $41.1

D. $205.5

29) The arcade changes the price shown below for game tokens. Using the pattern shown, what is the price for 5 tokens?

A. $2.00

B. $4.00

C. $5.00

D. $80.00

Tokens	Amount
20	$8.00
40	$16.00
60	$24.00
80	$32.00
100	$40.00

30) Suzy's class used $\frac{5}{8}$ of the cafeteria trays for their science display. There are 72 trays in all. How many trays were used?

A. 9

B. 27

C. 45

D. 360

Common Core Math Workbook

31) There are 2.2046 kilograms in a pound. How many kilograms are in 2 pounds?

   A. 4.4046

   B. 4.4092

   C. 4.4146

   D. 4.4192

32) Which of the following angles is acute?

   A. 80 Degrees

   B. 180 Degrees

   C. 90 Degrees

   D. 125 Degrees

33) A water tank contains 841 liters of water. A leak in the tank causes 1.5 liters to drip out each hour. How much water will the tank contain after 2 days?

   A. 838 liters

   B. 805 liters

   C. 793 liters

   D. 769 liters

34) If a secretory paid $43.50 for 6 rolls of film, how much did she pay for each roll?

   A. $7.25

   B. $37.50

   C. $72.50

   D. $261.

35) The perimeter of the trapezoid below is 29. What is its area?

A. 29 cm

B. 49 cm

C. 64 cm

D. 67.5 cm

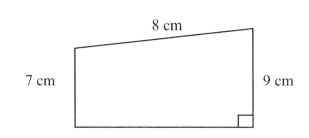

36) Joseph's first planter is 6 feet long and 2 feet wide and 5 feet height. The container is filled with soil to a height of 3 feet in the planter. What is the volume of soil in the planter?

A. 60 cubic cm

B. 96 cubic cm

C. 24 cubic cm

D. 36 cubic cm

37) The stem and leaf plot shows the numbers of miles run by each of runners. What is the difference between the number of runners who ran fewer than 30 miles and the number of runners who ran more than 54 miles?

A. 24

B. 13

C. 5

D. 4

Stem	Leaf
0	9
1	3 7 9
2	0 2 2 2 5
3	1 1 5 9
4	2 4 6
6	1 5 8
7	0

38) Linda will order ice-cream for her party. The relationship between S, the number of the scoops she will order, and C, the total costs of ice-cream she will pay, can be represented by the equation $C = 9S$.

Which table contain only values that represent the equation?

A. **Invoice**

Number of Scoops (S)	Cost (C)
5	$45
7	$63
12	$108
15	$145

B. **Invoice**

Number of Scoops (S)	Cost (C)
5	$45
7	$63
9	$108
15	$135

C. **Invoice**

Number of Scoops (S)	Cost (C)
1	$10
7	$16
11	$20
15	$24

D. **Invoice**

Number of Scoops (S)	Cost (C)
1	$9
7	$18
11	$27
15	$36

39) A cube has a volume of 64 cubic units. How many unit cubes will fit along one side of the cube if there are no gaps on overlaps?

A. 4

B. 8

C. 16

D. 64

40) Figure M and figure N are congruent. What is the length of side b in figure N?

A. 2

B. 3

C. 4

D. 5

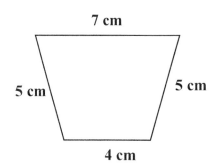
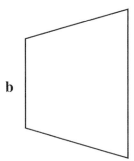

Answers and Explanations

Answer Key

Now, it's time to review your results to see where you went wrong and what areas you need to improve!

Common Core Math Practice Tests

Practice Test 1

1	B	16	D	31	D
2	C	17	A	32	C
3	C	18	D	33	A
4	A	19	A	34	C
5	B	20	C	35	D
6	D	21	D	36	B
7	B	22	D	37	D
8	D	23	A	38	C
9	C	24	B	39	C
10	A	25	C	40	A
11	C	26	B		
12	B	27	D		
13	C	28	C		
14	B	29	B		
15	D	30	A		

Practice Test 2

1	C	16	A	31	B
2	A	17	C	32	A
3	D	18	B	33	C
4	D	19	D	34	A
5	A	20	C	35	B
6	B	21	D	36	D
7	C	22	D	37	C
8	D	23	B	38	B
9	D	24	B	39	A
10	B	25	C	40	C
11	C	26	D		
12	D	27	D		
13	C	28	A		
14	B	29	A		
15	B	30	C		

Common Core Math Workbook

Common Core Practice Test 1
Answers and Explanations

1) Answer: B

The product of 7 and 12 is (7×12)

The difference of 19 and 8: $(19 - 8)$

Then, the sum: $(7 \times 12) + (19 - 8)$

2) Answer: C

Multiplication is the inverse of division and you can use multiplication to check your division answer (Dividend÷Divisor =Quotient; Then, Quotient × Divisor = Dividend).

$405 \div 15 = \Box \rightarrow \Box \times 15 = 405$

3) Answer: C

$\frac{1}{6} \div 18 = \frac{1}{6} \times \frac{1}{18} = \frac{1}{108}$

4) Answer: A

The bag of dimes should be divided by all students: $646 \div 34 = 19$ dimes

5) Answer: B

$6 \times 1.29 = \$7.74$

6) Answer: D

Associative property of multiplication: when multiplying three or more real numbers, the product is always the same regardless of their regrouping. By 'grouping' we mean how you use parenthesis. (a×b) ×c=a×(b×c)

7) Answer: B

To round a number to the nearest tenth, look at the next place value to the right (the hundredths). If it is 4 or less, just remove all the digits to the right. If it is 5 or greater, add 1 to the digit in the tenths place, and then remove all the digits to the right.

8) Answer: D

Each time spend $15 and, t times, means multiplication, then we need to use multiplication. The answer is: 15t

WWW.MathNotion.com

9) **Answer: C**

$4{,}716 \times 400 = 4{,}716 \times 40 \times 10$, then $\frac{4{,}716 \times 40 \times 10}{10} = 4{,}716 \times 40$

10) **Answer: A**

2 Shoes: $2 \times 35 = 70$; 25% of 70: $0.25 \times 70 = \$17.5$

3 Belts: $3 \times 10 = 30$; 5% of 30: $0.05 \times 30 = \$1.5$

$\$17.5 + \$1.5 = \$19$ the customer saved

11) **Answer: C**

The **PEMDAS Rule** (an acronym for "Parenthesis, Exponents, Multiplication, Division, Addition, Subtraction") is a set of rules that prioritize the order of calculations, that is, which operation to perform first. Parenthesis in math are used to group important things together, so you always do them first.

12) **Answer: B**

$20 \div 8 = \frac{20}{8} = 2\frac{4}{8} = 2\frac{1}{2}$ minutes

13) **Answer: C**

The sum of the measure of the three interior angles, in every triangle, is 180°.

A right triangle is a type of triangle that has one angle that measures 90°, then:

$90° + \alpha + \beta = 180° \rightarrow \alpha + \beta = 180° - 90° = 90°$

14) **Answer: B**

To multiply fractions is following the 3 steps: Multiply the numerators, Multiply the denominators, and Simplify the resulting fraction.

$\frac{7}{8} \times \frac{4}{9} = (\frac{1}{8} \times \frac{7}{1}) \times \frac{4}{9} = \frac{1}{8} \times (7 \times \frac{4}{9}) = \frac{7 \times \frac{4}{9}}{8}$

15) **Answer: D**

$7.5 \div 6 = 1.25$

16) **Answer: D**

Equivalent Fractions have the same value, even though they may look different. Because when you multiply or divide both the top and bottom by the same number, the fraction keeps its value. Then, $\frac{5}{3} = \frac{200}{?} \rightarrow \frac{5 \times 40}{3 \times 40} = \frac{200}{120}$

Common Core Math Workbook

17) Answer: A

20 feet long wire is cut into $\frac{4}{5}$ equal parts. Therefore, 20 should be divided by $\frac{4}{5}$

$20 \div \frac{4}{5} = 20 \times \frac{5}{4} = 25$

18) Answer: D

The pattern is: $57°, 61°, 65°$ (Add $4°$).

11:00A.M.−9:00A.M.=2:00 hours, and change to the half hours is $2 \times 2 = 4$, and use the pattern: $4 \times 4 = 16$; then, $16° + 65° = 81°$

19) Answer: A

Use volume of rectangle prism. V = width × length × heigth

$\Rightarrow 168 = 7 \times L \times 4 \Rightarrow 168 = 28L \Rightarrow L = \frac{168}{28} = 6$ uint-cubes

20) Answer: C

$1,062 \times 7 = 7,434$

21) Answer: D

Plug in 6 for m in the equations then, use order of operation rule (PEMDAS):

A. $6 + 12 \div m = 6 + 12 \div 6 = 6 + 2 = 8 \to 8 \neq 3$

B. $12 \div m + 6 = 12 \div 6 + 6 = 2 + 6 = 8 \to 8 \neq 1$

C. $18 + 6 \div m = 16 + 6 \div 6 = 18 + 1 = 19 \to 19 \neq 4$

D. $12 \div m + 2 = 12 \div 6 + 2 = 2 + 2 = 4 \to 4 = 4$

22) Answer: D

1 m=100 cm $\Rightarrow$ 11 m = 1100 cm

23) Answer: A

We can use an odd number as examples. If k is odd $2k$ is even (k=3, 6), and $2k \pm 1$ is odd (k=3,7or 5), and $k − 2$, is odd (k=3, 1). then, if k is odd, $k − 1$ is even (k=3, 2).

24) Answer: B

5 feet long ribbon is cut into 8 equal parts. Therefore, 5 should be divided by 8

25) Answer: C

$31.40 \div 5 = \$6.28$

26) Answer: B

We have two rectangular prisms.

Use volume of rectangle prism. V = width × length × heigth

$V_1 = 6 \times 2 \times 2 = 24$ cubic units

$V_2 = 8 \times 4 \times 2 = 64$ cubic units

$V = V_1 + V_2 = 24 + 64 = 88$ cubic unit

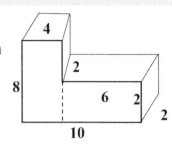

27) Answer: D

To determine the location of point D, the last vertex (corner) of the parallelogram (a four-sided figure with two pairs of parallel sides), you should have found a point exactly three units away (or 6 grid squares that are each 0.5 units long) from point C on the same horizontal grid line. Then, the ordered pairs that represents this location are (6.5,1.5) by right side and (0.5,1.5) by left side.

28) Answer: C

Composite numbers are a whole number that is a multiple of at least two numbers other than itself and 1. They have factors. $57 = 3 \times 19$.

29) Answer: B

First convert mL to L, then order from greatest:

521 mL=0.521 L;   5,215 mL=5.215 L,   6,001 mL=6.001 L

6.001 L, 6 L, 5.215 L, 5.21 L, 0.521 L

6001 mL, 6 L, 5215 mL, 5.21 L, 521 mL

30) Answer: A

$6\frac{2}{3} + 4\frac{5}{6} = 6\frac{4}{6} + 4\frac{5}{6} = 10\frac{4+5}{6} = 10\frac{9}{6} = 11\frac{3}{6} = 11\frac{1}{2}$

$12 - 11\frac{1}{2} = \frac{1}{2}$ Pizza

31) Answer: D

$12 \times 3 = 36$ tickets and each ticket cost 1.50: $36 \times \$1.50 = \$54$

Total cost: $\$54 + \$5 = \$59$

Common Core Math Workbook

32) Answer: C

To determine the perimeter (distance around the outside) of the hexagon, you should have added all of the side lengths.

$3.8 + 3.8 + 4.9 + 3.8 + 3.8 + 4.9 = 25$ in.

33) Answer: A

Percent means "out of 100." Or, Use percent formula: $\text{percent} = \frac{\text{part}}{\text{whole}} \times 100$

Whole $= 14 + 19 + 25 + 42 = 100$,

The nickels are 25, means: $\frac{25}{100}$ or 25%

34) Answer: C

Use percent formula: $\text{part} = \frac{\text{percent}}{100} \times \text{whole}$

Part $= \frac{80}{100} \times 25 = \frac{80 \times 25}{100} = \frac{2,000}{100} = 20$ students,

$20 \times 2 = 40$ crullers

35) Answer: D

Multiplying or dividing the numerator and denominator of a fraction by the same number will produce an equivalent fraction. $\frac{3}{7} \times \frac{4}{4} = \frac{12}{28}$

36) Answer: B

He wants to buy 2 cookies for 6 people: $2 \times 6 = 12$ cookies

Each cookie cost 60 cents, then, we need to have 12 of ¢60 or 12×60.

37) Answer: D

Perimeter is $P = 2l + 2w$, where l is the length and w is the width

$p = 2(2.5) + 2(1.6) = 5 + 3.2 = 8.2$ in.

38) Answer: C

We need to find the relation between input and output.

Input: $(5 - 1) = 4$, Output: $(14 - 2) = 12$, and the relation is 3 times of 4 is 12

Input, $(50 - 29) = 21$, use the relation: $3 \times 21 = 63$, then:

Output, $C - 86 = 63 \rightarrow C = 63 + 86 = 149$

WWW.MathNotion.com

Common Core Math Workbook

39) Answer: C

$19.63 + 42.12 + 128.66 = 190.41$

40) Answer: A

The sales are: 8, 12, 20, 10, 18, and 22

Write the numbers in order: 8, 10, 12, 18, 20, 22

Median: $\frac{12+18}{2} = \frac{30}{2} = 15$

Mean: $\frac{\text{sum of terms}}{\text{number of terms}} = \frac{8+10+12+18+20+22}{6} = \frac{90}{6} = 15$

Median − Mean = $15 - 15 = 0$

Common Core Practice Test 2

Answers and Explanations

1) Answer: C

Place Value Chart								
Millions			Thousands			Ones		
Hundreds	Tens	Ones	Hundreds	Tens	Ones	Hundreds	Tens	Ones
	2	3	0	9	0	5	6	2

In the number 23,090,562 the digit 0 is in the (one) thousands place.

2) Answer: A

You should have used the order of operations, or PEMDAS. 1. Operations contained in Parentheses or brackets, 2. Exponents 3. Multiplication/Division from left to right, and 4. Addition/Subtraction from left to right.

$4[4.5 - 2(1.2)] = 4[4.5 - 2.4] = 4[2.1] = 8.4$

3) Answer: D

Multiplying or dividing the numerator and denominator of a fraction by the same number will produce an equivalent fraction. $\frac{72}{108}$ divided by 9 = $\frac{8}{12}$ divided by 4 = $\frac{2}{3}$

4) Answer: D

The distributive property lets you multiply a sum by multiplying each addend separately and then add the products.

$53 \times 24 = 53 \times (20 + 4) = (53 \times 20) + (53 \times 4)$

5) Answer: A

The place value of 6 in 8.64 is tenth

The place value of 6 in 6.48 is ones

To convert tenth place to ones we need to multiply by 10.

6) Answer: B

$92.53 \div 10 = 9.523$

7) Answer: C

When converting word names to standard form. The words "million" and "thousand" tell you which periods the digits are in.

8) Answer: D

1 L = 1,000 mL; Convert 0.5 L to mL: $0.5 \times 1,000 = 500$ mL

9) Answer: D

Add decimals: 7.65+2.8=10.45

Convert decimal to fraction: $10.45 = 10\frac{45}{100}$, then simplify fraction (by 5): $10\frac{9}{20}$

10) Answer: B

$26.4 \div 4 = 6.6$ in one hour

$6.6 \div 60 = 0.11$ in one minute,

$5 \times 0.11 = 0.55$ mile

11) Answer: C

$2 \times 36.95 + 5 \times 23.95 = 73.90 + 119.75 = 193.65$

12) Answer: D

$18 \times 864 = 15,552$

13) Answer: C

Compare the fractions: $\frac{3}{4} > \frac{1}{2} > \frac{1}{3} > \frac{3}{10}$

Therefore, $\frac{3}{10}$ is the least fraction and related to stamp 3

14) Answer: B

You should have first determined that her weekly gross income (income before paying taxes) is $22.5 \times \$12 = \$270$. But, her weekly net income (income after paying taxes) would be less than her weekly gross income of $270.

15) Answer: B

$9.80 \div 1.40 = 7$

16) Answer: A

$194.38 + 478.56 = 672.94$, then subtract from $800: $800 - 672.94 = 127.06$

Common Core Math Workbook

17) Answer: C

Plug in $n = 12$ and $n = 14$ in each equation.

A. $\frac{n}{3} \to \frac{12}{3} = 4$ and $\frac{14}{3} \neq 5$

B. $n - 9 \to 12 - 9 \neq 4$

C. $\frac{n}{2} - 2 \to \frac{12}{2} - 2 = 6 - 2 = 4$ and $\frac{14}{2} - 2 = 7 - 2 = 5$ Bingo!

D. $\frac{n}{2} + 2 \to \frac{12}{2} + 2 = 6 + 2 \neq 4$

18) Answer: B

All student is 10, and 4 of them made 5 and more, then $\frac{4}{10}$ and simplify: $\frac{2}{5}$

19) Answer: D

Use percent formula: $percent = \frac{part}{whole} \times 100$

$Percent = \frac{28}{70} \times 100 = 40\%$ coin lands on head. 100%−40%=60% land on tail.

20) Answer: C

Area of parallelogram: $A = b.h = 32 \times 25 = 800 \ cm^2$

Area of triangle: $A = \frac{1}{2} b.h = \frac{1}{2} \times 66 \times 25 = 825 \ cm^2$

Area of triangle− Area of parallelogram $= 825 - 800 = 25 \ cm^2$

21) Answer: D

Compare the fractions:

$\frac{3}{5} > \frac{1}{2}$; $\frac{2}{3} > \frac{1}{2}$; $\frac{1}{6} < \frac{1}{5}$; $\frac{1}{3} > \frac{1}{5}$, and $\frac{1}{3} < \frac{1}{2}$

22) Answer: D

Convert fraction to decimal: $\frac{13}{25} \times 100 = \frac{13 \times 100}{25}$ both numerator and denominator divided by 25: $\frac{13 \times 4}{1} = 52\%$

23) Answer: B

$Probability = \frac{number \ of \ desired \ outcomes}{number \ of \ total \ outcomes} = \frac{5}{3+2+5+4+6} = \frac{5}{20} = \frac{1}{4}$

24) Answer: B

Quadrilaterals are four sided polygons, with four vertexes, whose total interior angles add up to 360 degrees.

$80° + 110° + 90° = 280°; 360° - 280° = 80°$

25) Answer: C

Factor of 74: $(1, 2, 6, 12, 37, 74)$

Factor of 42: $(1, 2, 3, 6, 7, 14, 42)$

Greatest Common Factor is: 6

26) Answer: D

when decimals are compared start with tenths place and then hundredths place, etc. If one decimal has a higher number in the tenths place, then it is larger than a decimal with fewer tenths. If the tenths are equal compare the hundredths, then the thousandths etc.

$0.34 < 0.43$ is correct.

27) Answer: D

Convert decimal to percent by multiply 100: $0.4 \times 100 = 40\%$

28) Answer: A

$3 \times \$12.60 = \$37.80$, and $2 \times \$28.5 = \$57$

$\$37.80 + \$57 = \$94.80$

29) Answer: A

The rule is $y = 0.40\ x$, when y is output (amount) and x is input (Token).

Amount of 5 token is: $Amount = 0.4 \times 5 = \$2.00$

30) Answer: C

$\frac{5}{8} \times 72 = \frac{5 \times 72}{8} = \frac{5 \times 72^{\ 9}}{8_{\ 1}} = 45$

31) Answer: B

$2 \times 2.2046 = 4.4092$

32) Answer: A

An acute angle is an angle of less than $90°$.

From the options provided, only option A (80 degrees) is an acute angle.

33) Answer: C

One day equals to 24 hours, and 2 days: $2 \times 24 = 48$ hours

$48 \times 1.5 = 72$, $841 - 72 = 769$ liters

34) Answer: A

$43.50 \div 6 = \$7.25$ each roll

35) Answer: B

First, find the missing side of the trapezoid. The perimeter of the trapezoid below is 70.

Therefore, the missing side of the trapezoid (its height) is:

$29 - (5 + 8 + 9) = 29 - 22 = 7$

Area of a trapezoid: A = $\frac{1}{2}$ h (b1 + b2) = $\frac{1}{2}$ (7) (9 + 5) = 49 cm

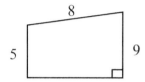

36) Answer: D

Use volume of rectangle prism formula.

$V = length \times width \times height \Rightarrow V = 2 \times 6 \times 3 = 36$ cubic cm

37) Answer: C

There are 9 values on the stem and leaf plot that are less than 30 (9, 13, 17, 19, 20, 22, 22, 22, and 25) and 4 values on the stem and leaf plot that are greater than 54 (61, 65, 68, and 70). Then subtracted 4 from 9, resulting in a difference of 5.

38) Answer: B

To determine each C-value in the table, we should have multiplied 9 by each S-value: ($5 \times 9 = 45$; $7 \times 9 = 63$; $12 \times 9 = 108$; and $15 \times 9 = 135$).

39) Answer: A

Cube is a three-dimensional shape that has equal width, height, and length measurements. If a cube has side length "S" then volume is: $V = S \times S \times S = S^3$

$V = S \times S \times S \rightarrow 64 = S \times S \times S$, plug in the small number of options provided to find the length of side. If $S = 4$ $V = 4 \times 4 \times 4 = 64$, means $S = 4$ cubes

40) Answer: C

When shapes are congruent, all corresponding sides and angles are also congruent. The same shape and size, but we can flip, slide or turn. Then, $b = 4cm$

"End"

Printed in Great Britain
by Amazon